# TOWARD A
# BEHAVIORAL ECOLOGY
# OF LITHIC TECHNOLOGY

TOWARD A
# BEHAVIORAL ECOLOGY
# OF LITHIC TECHNOLOGY
## CASES FROM PALEOINDIAN ARCHAEOLOGY

TODD A. SUROVELL

*The University of Arizona Press   Tucson*

### THE UNIVERSITY OF ARIZONA PRESS

© 2009 The Arizona Board of Regents
All rights reserved
First issued as a paperback edition 2011

www.uapress.arizona.edu

Library of Congress Cataloging-in-Publication Data
Surovell, Todd A., 1973–
Toward a behavioral ecology of lithic technology :
cases from Paleoindian archaeology / Todd A. Surovell.
p. cm.
Includes bibliographical references and index.
ISBN 978-0-8165-2810-3 (cloth : alk. paper)
ISBN 978-0-8165-0738-2 (pbk. : alk. paper)
1. Paleo-Indians—Great Plains. 2. Paleo-Indians—
Rocky Mountains. 3. Social archaeology—Great Plains.
4. Social archaeology—Rocky Mountains. 5. Human
ecology—Great Plains—History. 6. Human ecology—
Rocky Mountains—History. 7. Stone implements—Great
Plains. 8. Stone implements—Rocky Mountains. 9. Great
Plains—Antiquities. 10. Rocky Mountains—Antiquities.
I. Title.
E78.G73S895           2009
978.004997—dc22      2009014334

Manufactured in the United States of America on acid-free,
archival-quality paper and processed chlorine free.

16  15  14  13  12  11    8  7  6  5  4  3

# Contents

List of Illustrations   vii
Preface   xiii
Acknowledgments   xvii

1  Toward a Behavioral Ecology of Lithic Technology   1

2  Late Pleistocene Foragers of the Northern Plains and Rocky Mountains   23

3  Occupation Span and Residential Mobility   58

4  The Reoccupation Problem   99

5  Stone Age Supply-Side Economics   110

6  Bifaces, and So On: Modeling the Design of Tools and Toolkits   142

7  On the Optimal Production of Trash   177

8  Mathematics, Lithic Technology, and Paleoindians   213

Appendix: Site Occupancy and Camp Area   233
Notes   237
References   241
Index   267

# Illustrations

**Figures**

1.1. Caveman calculus  3
1.2. Metcalfe and Barlow field processing model  14
1.3. Kuhn model of tool blank size vs. utility:mass  17
1.4. Brantingham and Kuhn's formal model of Levallois reduction  18
2.1. Map of geographic range of Folsom points  26
2.2. Map of geographic range of Goshen projectile points  28
2.3. Site-averaged charcoal or bone collagen radiocarbon dates for seven Folsom sites  31
2.4. Site-averaged charcoal radiocarbon dates for three Goshen sites  32
2.5. Aerial view of Barger Gulch, Locality B  46
2.6. A sample of artifacts from Barger Gulch, Locality B  47
2.7. View of Upper Twin Mountain Goshen bison kill, Middle Park, Colorado  48
2.8. A sample of artifacts from Upper Twin Mountain  49
2.9. View of Krmpotich Folsom site, Sweetwater County, Wyoming  50
2.10. A sample of artifacts from the Krmpotich site  51
2.11. Aerial View of Agate Basin site  52
2.12. A sample of artifacts from Agate Basin, Area 2, Folsom component  53
2.13. View of Carter/Kerr-McGee site, Natrona County, Wyoming  54
2.14. A sample of artifacts from Carter/Kerr-McGee Folsom component  55
3.1. Gallivan and Lightfoot and Jewett models for distinguishing single vs. multiple and short- vs. long-term occupations  65

3.2. Cumulative number of type 1 and type 2 artifacts discarded as a function of time   74
3.3. Cumulative number of class a and class b artifacts discarded as a function of time   76
3.4. Ratio of class b to class a artifacts vs. time   77
3.5. Modeled relationships between size of initial transported toolkit, optimal size of toolkit, and artifact use-life on ratio of local: nonlocal raw materials   79
3.6. Modeled relationship between time and ratio of debitage: transported tools   84
3.7. Local:nonlocal raw materials and debitage:nonlocal tools for simulated assemblages   85
3.8. Local:nonlocal raw materials vs. debitage:nonlocal tools for the five study sites   86
3.9. Local:nonlocal raw materials vs. debitage:nonlocal tools for enlarged sample of sites   89
3.10. Local:nonlocal raw materials vs. debitage:nonlocal tools for Puntutjarpa rockshelter, Australia   90
3.11. Australian Holocene lake levels vs. local:nonlocal raw material and debitage:nonlocal tools from Puntutjarpa rockshelter   95
4.1. A model for distinguishing between single and multiple occupations   100
4.2. Artifact density vs. local:nonlocal raw materials, debitage: nonlocal tools, and occupation span index for Puntutjarpa rockshelter   102
4.3. Artifact density vs. occupation span index for Puntutjarpa rockshelter   103
4.4. Artifact density vs. occupation span index for study sample   105
4.5. Simulated relationship between number of occupations and likelihood and number of spatially overlapping occupations   109
5.1. U.S. federal budget deficit or surplus for fiscal years 1962–2007   111
5.2. Modeled actual and average rate of consumption of lithic raw materials for an individual over span of an occupation   114
5.3. Modeled probability of a shortfall as a function of lithic surplus size   116

List of Illustrations    ix

5.4.  Modeled optimum surplus sizes for various occupation spans   117
5.5.  Modeled relationship between occupation span and optimal surplus size   119
5.6.  Modeled relationship between proportionality constant and optimal surplus size   119
5.7.  Predicted relationship between occupation span and proportion of assemblage composed of surplus lithic raw material   120
5.8.  Occupation span index vs. percent surplus raw material for Puntutjarpa rockshelter   123
5.9.  Two measures of occupation span vs. percent surplus based on artifact count for study sample   127
5.10. Two measures of occupation span vs. percent surplus based on artifact mass for study sample   127
5.11. A model of raw material procurement and the costs of direct and embedded procurement   129
5.12. Modeled costs of direct and embedded procurement vs. distance to a single lithic source   130
5.13. Maximum foraging distance:distance to a single raw material source vs. direct:embedded procurement ratio   131
5.14. Number of lithic raw material sources vs. direct:embedded procurement ratio for multiple-source raw material configurations   132
5.15. Distance to closest raw material source vs. cost of direct and embedded procurement for multiple-source raw material configurations   133
5.16. Known raw material source areas within 3 km of Barger Gulch and Krmpotich sites   134
5.17. Maximum foraging distance vs. cost of direct:embedded procurement for Barger Gulch and Krmpotich   136
5.18. Artifact density vs. surplus size for study sample with Barger Gulch adjusted by estimation of relative costs of direct and embedded raw material procurement   137
6.1.  Schematic representation of flake tool model   144
6.2.  Modeled flake tool length vs. transport efficiency   145
6.3.  Modeled flake thickness:length vs. transport efficiency   146

6.4. Histogram of endscraper lengths, expressed as a multiple of the shortest scraper   147
6.5. Transport efficiency vs. length for cores   151
6.6. Schematic representation of biface model   158
6.7. Modeled relationships between length, thickness, and minimum usable size vs. transport efficiency for bifacial tools   160
6.8. Modeled relationships between length and thickness vs. transport efficiency for bifacial cores   161
6.9. Modeled relationship between length and transport efficiency for bifacial core tools   163
6.10. Schematic of predicted bifurcation in bifacial core tool design   166
6.11. Property space map of bifacial core tool model   167
6.12. Transport efficiency of bifacial core tools holding volume constant   168
7.1. Four measures of occupation span vs. core reduction index for study sample   185
7.2. Three hypotheses to explain changing ratios of core reduction to bifacial thinning debris   186
7.3. Occupation span index vs. core density for study sample   191
7.4. Marginal theorem patch choice model as applied to stone tool use and reduction   193
7.5. Percentage of retouch on flake tools by edge unit for sites of study sample   198
7.6. Occupation span index vs. retouch intensity index for study sample   199
7.7. Core mass density vs. retouch intensity index for study sample   199
7.8. A conceptual model of tool design and flakes produced in core reduction   202
7.9. Number of flakes produced in tool manufacture vs. mean production efficiency   204
7.10. Core mass density vs. core reduction flakes:flake tools and bifacial thinning flakes:bifaces   207
8.1. Occupation span index vs. ratio of core reduction to bifacial thinning flakes for the study sample in comparison to the Hanson site   228

List of Illustrations    xi

## Tables

3.1. Artifact raw material frequencies, type frequencies, and local: nonlocal raw material ratios by site for artifacts larger than 2 cm    81
3.2. Raw material totals and debitage and nonlocal tool counts tabulated by site for artifacts larger than 2 cm    85
3.3. Ratios of local:nonlocal raw material and debitage:nonlocal tools for artifacts in study sample and from enlarged sample    88
3.4. Artifact density, local:nonlocal raw materials, debitage:nonlocal tools, and occupation span index by stratigraphic level for Puntutjarpa rockshelter    93
4.1. Occupation span index and horizontal artifact densities for sites of study sample for artifacts larger than 2 cm    104
5.1. Typological divisions of artifact types from Puntutjarpa into surplus and consumed lithic raw materials    122
5.2. Counts of surplus and consumed artifacts from Puntutjarpa shelter    124
5.3. Artifact counts for five sites of study sample by type and length    125
5.4. Total artifact mass for five sites of study sample by type and length    126
6.1. Technological attributes of tools and debitage    149
6.2. Counts of cores by site and distance to source area    154
6.3. Technological attributes of local and nonlocal debitage larger than 2 cm    156
6.4. Reduction stage for nonlocal debitage larger than 2 cm    157
7.1. Technological attributes of debitage by site for all artifacts analyzed independently    182
7.2. Correlation matrices of debitage attributes for two samples    182
7.3. Standardized debitage attribute values, core reduction index, and measures of occupation span by site    184
7.4. Core density and core mass density for sites of study sample    190
7.5. Percentage of tool edge units exhibiting retouch by site    196
7.6. Measures of average tool use-life by site    197
7.7. Counts of debitage, tools, and bifaces by site    206
8.1. Original and revised relative occupation spans for sites of study sample using artifacts larger than 2 cm    223

# Preface

I LIKE MATH, but I'm not very good at it. To some who read this book, this statement may seem like unnecessary modesty or wholly inaccurate, but to those who are much more math savvy than I, my clumsy notation and limited abilities will be readily apparent. I like expressing cause and effect mathematically because for me it leads to clarity in explanation. That said, I recognize that anthropology and archaeology are dominated by individuals who by and large do not like math, some who fear math, and some who cringe at the idea of expressing human behavior as a series of abstract variables linked by mathematical operators. Yet, I think most people recognize at least some value in the approach. It works for me, and I am very aware that it may not work for you. I am also well aware of the feeling of losing interest in reading something when you get to the part of a text that is peppered with equations. Shamefully, I must admit that my eyes occasionally glaze over when I see other people's equations. I wrote this book with this in mind. The following text is, in places, peppered with equations, but for the most part they are not complex. And I do not leave them to speak for themselves. I have made a point to explain equations in simple terms for those readers who are not accustomed to viewing the world mathematically.

A love of mathematics for me came early in life. During the course of my math education, however, I became increasingly apathetic. Analytical geometry, trigonometry, functions, and calculus increasingly seemed like wastes of my time. Now, I think a basic working knowledge of at least trigonometry is a must for an archaeologist. I came back to math in graduate school. Wanting to study Paleoindian archaeology at the University of Arizona, I was disappointed in my inability to access actual Paleoindian collections. This changed later, but early on I turned to another skill in my toolkit—computer programming. Since I could not study the archaeological record directly (in the way I wanted to), I decided I could create a fantasy archaeological world in silicon circuitry. Why not?

The closest thing to a "Eureka!" moment that I have ever experienced occurred during my early and even clumsier efforts at modeling in graduate school. I read the paper "New Perspectives on the Clovis vs. Pre-Clovis Controversy" by David Whitley and Ronald Dorn (1993), published in *American Antiquity*. I was intrigued by the beginning of the paper, which suggested that the archaeological record of Clovis presented a paradox. Clovis seemingly appeared from coast to coast within a matter of centuries. In order to accomplish colonization at such a rapid pace, Clovis folks would have had to be prodigious reproducers. Yet, Clovis peoples were renowned for high mobility evidenced by regular transport of lithic raw materials over huge distances. Conventional wisdom held that "high mobility" and high fertility were incompatible because foragers must carry children everywhere they go. The implication: Clovis was not first.

I set out to demonstrate that Whitley and Dorn were correct by mathematically modeling the relationship between the organization of mobility and the cost of raising children for hunter-gatherers. Because the math was too complex for me at the time, I wrote a program to do the calculations for me. When the computer gave me the results of its calculations, they were exactly the opposite of what I was expecting. The computer was telling me that frequent residential mobility was quite compatible with high fertility because moving camps frequently minimizes the costs of moving kids around. My first reaction was, "Where did I screw up?" After much examining of code, I realized it was not my programming that was flawed but my intuition. Intuitive qualitative modeling can be a very useful thing, but it can also lead us to incorrect conclusions. Whether I was right or wrong about early Paleoindians (Surovell 2000), it was likely that moment plus a few more that pushed me in the direction of mathematical modeling.

When I set out to begin this study, my skills in calculus were sorely lacking, and I needed a refresher. I found comfort in the book *Calculus Made Easy* by Silvanus P. Thompson, first published in 1910. This book, which has been reprinted many times, has the lengthy but telling subtitle: *Being a Very-Simplest Introduction to Those Beautiful Methods of Reckoning Which Are Generally Called by the Terrifying Names of the Differential*

*Calculus and the Integral Calculus.* Thompson's subtitle captures two important points. First, he recognizes the beauty of mathematical formulations. To represent any process or system mathematically is a beautiful thing. Mathematical models are simple abstractions of complex systems that bring clarity to explanation of observable properties of the world. Of course, simple is a relative term. What is "simple" to one may be "terrifying" to another (the second point of the subtitle). To those who also feel the need for a refresher course in calculus, I recommend this book highly. Thompson is also well aware of a math-anxious audience and attempts to provide solace early on: "Considering how many fools can calculate, it is surprising that it should be thought either a difficult or tedious task for any other fool to learn how to master the same tricks. . . . Being a remarkably stupid fellow, I have had to unteach myself the difficulties, and now beg to present to my fellow fools the parts that are not hard. . . . What one fool can do, another can" (Thompson and Gardner 1998:38).

# Acknowledgments

THROUGHOUT THE COURSE of this study, numerous people have shared their thoughts with me, given me advice, and generally supported my efforts. Those who have been most influential include Nicole Waguespack, Steve Kuhn, Mary Stiner, Vance Haynes, and Jeff Brantingham. I am grateful to David Meltzer, who read and commented on an early draft of this study. Many others have helped in numerous ways, including George Frison, Bob Kelly, Mary Lou Larson, Marcel Kornfeld, Chuck Reher, Mark Miller, Rick Weathermon, Adam Graves, Allison Byrnes, Joseph Daniele, John Laughlin, Melissa Daniele, Sage Wall, Mike Peterson, Judy Brown, and Rhoda Owen Lewis. I was able to perform this research in large part thanks to an Emil W. Haury Fellowship provided by the Department of Anthropology at the University of Arizona. Also, my analysis of the assemblage from Barger Gulch Locality B would not have been possible if not for collaboration with Nicole Waguespack, Marcel Kornfeld, and George Frison. Our work at Barger Gulch has been supported by grants from the National Science Foundation (#0450759), Colorado State Historical Fund (#01–02–122), the Emil W. Haury Fund for Archaeology, and the Colorado Bureau of Land Management. Many other individuals have aided us in our work in Middle Park, including Frank Rupp, Art and Roberta Bruchez and their family, Jim Chase, and Tony Smith. Nicole Waguespack has been my best friend and colleague through my academic career. More than anyone else, she taught me how to be an archaeologist. She makes difficult problems seem simple. She is my greatest fan and most thoughtful critic. I must admit that this entire study evolved from an insight she shared with me in a conversation seven years ago. For all she has done, I am incredibly grateful and can only begin to repay her deeds.

# TOWARD A BEHAVIORAL ECOLOGY OF LITHIC TECHNOLOGY

# 1

# Toward a Behavioral Ecology of Lithic Technology

IN HER EDITED volume *Time, Energy, and Stone Tools*, Robin Torrence (1989b:1) wrote: "Archaeologists have been notoriously poor at producing their own theories for behavior and have depended largely on borrowing from anthropology and ecology, with, it must be admitted, mixed results." What Torrence was lamenting, I believe, is something felt by many archaeologists. We have masterfully created a superabundance of data through more than a century of fieldwork, but to make sense of it we have relied heavily on models borrowed from other fields rather than building archaeological theory from the ground up. This is especially true for the most common artifact in the archaeological record—the lowly flake—or perhaps the "lonely" flake, which until recently received relatively little attention from lithic analysts. Modern humans and their hominid ancestors relied on chipped stone technology for well over two million years and colonized more than 99 percent of the Earth's habitable landmass doing so. Yet we have only a handful of informal models derived from ethnographic observation, experiments, engineering, and "common sense" to explain variability in archaeological lithic assemblages. For this reason, the study of lithic technology seems to have stalled.

The intent of this study is to begin to develop a formal theory of lithic technology. Although, as many researchers have noted, people who make and use stone tools are no longer extant (except in a very few limited circumstances), this in no way should hamper our ability to develop theory pertaining to lithic technology. We have not, for instance, allowed the lack of extant cave artists or governments commissioning pyramids to stifle efforts to theoretically explore these topics. Because the fundamental processes of making, using, and discarding stone tools are, at their very root, exercises in problem solving, we can ask what conditions favor certain technological solutions. Whether we are asking how wide the business end of a scraper should be or whether a flake should be saved or discarded, answers must be sought that extend beyond a case-by-case basis. One avenue for addressing these questions theoretically is formal mathematical modeling.

I define "formal model" as a model that is constructed mathematically, either built from mathematical expressions or algorithms (computer programs). Formal models have the advantage of having explicit predictions that must derive from their assumptions, something not true of informal models. As Kelly (1992:56) has noted, "At present, then, many interpretations of stone tool assemblages as indicators of mobility are subjective, intuitive, and sometimes contradictory." These contradictions often arise not from data themselves but from theoretical inadequacies, resulting from the development and application of models that occasionally lack a logical foundation because they are often constructed with faulty, or at least unsubstantiated premises, or because they do not explicitly state goals, currencies, or constraints. My intent is not to discredit all previous work based on informal models, nor to say that there is only one way to "do" archaeology. Formal mathematical models are but one of a myriad of tools that can be used to learn about the world. Instead, I intend to highlight the utility of formal mathematical models for a particular area of archaeological research.

Mathematical models, by their very nature, entail model causal relationships that have unambiguous predictions. In contrast to some early criticisms of the use of formal models of kin selection in anthropology (e.g., Sahlins 1976:44–45), formal mathematical modeling of human behavior does not assume that prehistoric humans were slide rule–toting Einsteins who moved from camp to camp calculating optimal solutions to foraging problems (Boone and Smith 1998; Dawkins 1989:291–292). After all, there is overwhelming evidence that all species of animals from invertebrates to vertebrates are optimizers (Alexander 1996; Krebs and Davies 1984), even though they are rarely if ever conscious of the underlying principles governing their behavior (fig. 1.1). Of course, there is a simple explanation of why animals are optimizers—behaviors that maximize fitness are favored by natural selection, a process in and of itself that rewards optimization (at least with respect to the number of surviving offspring) for a given set of environmental variables. Furthermore, there are many studies that show unequivocally that complex human subsistence behaviors can be explained with reference to very simple mathematical

FIGURE 1.1. Caveman calculus.

models from foraging theory (e.g., Belovsky 1987; Hawkes et al. 1982; O'Connell and Hawkes 1981; Smith 1991; Winterhalder 1981). Unfortunately, it seems as if solving equations for optima is often more difficult than putting the theory into practice on the ground. Humans who have made a living by foraging through trial and error over many generations seem to arrive at optimal or near-optimal solutions for many problem-solving tasks at hand. While the basic principles underlying the decision-making processes hunter-gatherers engage in may or may not be evident to them, the process of formal model building often illuminates many of the key variables and relationships between those variables that often are not self-evident.

As archaeology and anthropology are firmly grounded in the social sciences, I believe many archaeologists have an aversion to mathematics in large part because of the nature of our education. It is less taxing and more familiar to construct verbal arguments than to develop theory based on relationships between abstract alphabetic variables, often prefaced with Greek symbols. The historical trajectory of archaeological and anthropological hypothesis testing has favored the use of model building that relies essentially on verbal arguments constructed from what is often portrayed as logic but usually is only a series of causal relationships supported by "common sense" or received wisdom. Mathematical aversions aside, many anthropologists adhere to the belief that indiscriminate application of formal behavioral models to arthropods and your Uncle Arthur alike demeans the uniqueness of human behavior. For example, a few years back when I told one of my professors that I was having some success predicting variation in chipped stone assemblages using formal mathematical models, he responded, "That's depressing." Whether you find it depressing or enlightening that the same model can explain the foraging behavior of a praying mantis and a praying primate (e.g., Charnov 1976a; Hawkes et al. 1982), the fact is that humans do things purposefully, not randomly, and mathematical models are one way of clearly and explicitly exploring the factors governing human decision making. Finally, for those who feel that behavioral ecological approaches are limited to the material and economic domains, I refer them to Pascal's wager (Pascal 1670), which could be considered the first formal optimality model and concerns the decision of whether or not to believe in god-given specific costs, benefits, and constraints. (For a true appreciation of the scope of behavioral ecology in anthropology and archaeology, see Bird and O'Connell 2006 and Winterhalder and Smith 2000.)

The goal of this study is to begin to develop and test formal models of lithic technology for hunter-gatherers. This research is framed in terms of behavioral ecology wherein evolution by natural selection, played out within a given environment, is the ultimate causal determinant of human behavior. In particular, I explore the impacts of variation in forager mobility and raw material availability on technological variability in archaeological lithic assemblages. A second goal is to reconstruct patterns of Paleoindian mobility and land use in these regions. The study is based on the analysis of more than 45,000 chipped stone artifacts from five archaeological sites in Wyoming and Colorado falling within the Younger Dryas climatic episode (11,000–10,000 $^{14}$C yr BP). The study also includes lithic data compiled from published Paleoindian sites located throughout the western Great Plains and Rocky Mountains.

## Behavioral Ecology: A Brief Background

Behavioral ecology has its roots firmly planted in the biological sciences. With the related paradigm of sociobiology, behavioral ecology provides a rigorous framework for studying adaptation within an environmental context (Bird and O'Connell 2006; Smith 2000; Winterhalder and Smith 1992, 2000). Following Alcock (2001:9), behavioral ecology explores the relationship between behavior and the natural environment, while sociobiology studies behavior as it relates to the social environment. Obviously the separation between sociobiology and behavioral ecology is somewhat indistinct when based on a division of external factors affecting behavior into intraspecific interactions and all other environmental stimuli. Since both can reasonably be expected to shape behavior simultaneously (Giraldeau and Caraco 2000), perhaps the bipartite theoretical classification is unnecessary.

In archaeology and anthropology, behavioral ecology has a growing following (Winterhalder and Smith 2000:fig. 1), while very few researchers would admit to being sociobiologists. This likely arises from the harrowed history of sociobiology (Kitcher 1985; Wilson 1994; Winterhalder and Smith 1992:18–20, 2000:51). In many fields the term "sociobiology" is treated as a four-letter word. The notion that human behavioral adaptation can be explained with reference to our evolutionary history rather than our cultural history has meant to some that sociobiology could be used as a justification for rape, murder, genocide, discrimination, and many other unpleasantries of which the human animal is capable and prone to commit (Allen et al. 1975). However, the fact or even hypothesis that violent and unsavory behaviors may have natural explanations derived from our shared evolutionary history in no way translates to an argument that such behaviors are morally excusable (Alcock 2001). It does necessitate, however, that socially and politically relevant issues in the science of human behavior be approached with caution and that proclamations about "human nature" should not proceed without extreme certainty of their veracity (Kitcher 1985).

Often, the term "evolutionary ecology" is used synonymously with behavioral ecology, though some consider the latter to be a subset of the former (Bird and O'Connell 2006). Although models in behavioral ecology are ultimately founded on evolutionary principles, they generally do not examine evolutionary origins directly, nor do they directly monitor changes in gene frequencies through time (Bamforth 2002a), at least when applied to archaeological data. Because most models in behavioral ecology are not diachronic, it is somewhat of a misnomer to label them as "evolutionary." In this study, I am largely concerned with the near synchronic adaptations of individuals as they relate both to cultural and natural (nonhuman) environments, and for this reason, I have chosen to subsume this study under the rubric of behavioral ecology. Furthermore, most models in human behavioral ecology do not specify a mechanism of transmission (Bird and O'Connell 2006), whether biological or cultural, though the latter is certainly the dominant force for transmission of many behavioral phenotypes among humans (see Boyd and Richerson 1985). If acquired phenotypes are culturally transmitted among individuals, then calling human behavioral ecology "Darwinian" may be a misnomer as well. In this sense, "Lamarckian" may be a better descriptor.

Behavioral ecology is fundamentally about problem solving. To make a living, organisms must not only meet nutritional and physiological requirements; to be successful in an evolutionary sense, they must also maximize mating opportunities (and/or inclusive fitness). Organisms meet this multitude of needs through biological, behavioral, and, for humans (and other organisms), cultural adaptations. For individual organisms through time, selection will favor behaviors that maximize the number of surviving offspring. This will obviously apply directly to reproductive behaviors but should apply to other realms of the behavioral repertoire as well. A behavior, for example, that enhances the efficiency of food capture for an individual translates to better nutritional status, more time available to dedicate to reproductive activities, or more energy to devote to nurturing offspring, thus increasing their chances for survival. Therefore, the ultimate explanation for any behavior in behavioral ecology is thought to be its evolutionary history, even if the selective process is not studied directly.

This generalization should not be construed as a statement that all behaviors will ultimately be adaptive, in the sense of fertility maximization, for at least two reasons. First, as any organism is the cumulative product of its evolutionary history, it may retain adaptations that are no longer beneficial to its fitness and may even be detrimental. Evolution, in fact, guarantees that a lag will exist between environmental change and biological change, since evolution works only on generational time while environments can change rapidly. A well-known example of such a lag is the high incidence of diabetes in some Native American populations, resulting from physiologies that are not well equipped to deal with dietary changes that have become commonplace within only the last few generations (Weiss et al. 1984). Unlike human physiology, however, human behavior is very fluid and can adapt rapidly.

More importantly, the relationship between genes and behaviors is a complex one. Generally speaking, there is not a one-to-one correlation between genes and behavior; that is, there are no single genes that control single behaviors, such as a gene that would determine whether a person would be stingy or wasteful. Although the exact mechanisms remain obscure, behaviors are ultimately the product of the interaction of myriads of genes. In such a system, neutral or maladaptive behavioral traits are likely to evolve because increasing the frequency of certain beneficial genes could have some negative behavioral consequences or side effects, while still producing a net increase in fitness (Alcock 2001). This could explain the existence of some traits in humans, such as masturbation, suicide, and adoption of others' children that seem to have no adaptive value with respect to fitness.

Behavioral ecology, therefore, assumes that behavior is a product of natural selection, and while it generally should confer adaptive benefits, it occasionally will not. If so, then what ultimately governs behavior? Fundamentally, it is assumed that behavior is not random but instead is geared toward the optimization of some currency, usually framed in terms of a cost-benefit analysis. Behavior, then, can be construed as maximizing benefits or minimizing costs. Models in behavioral ecology may include both costs and benefits, or alternatively solely costs or benefits. Also, a model must specify a goal and a decision variable. The goal simply states the intent of the agent and is specified relative to a currency—for example, the maximization of the rate of energy captured or minimization of time spent foraging. The decision variable defines what aspect of behavior will be adjusted to meet this goal. Models in behavioral ecology are framed in terms of constraints, essentially assumed limits on behavior, determined by the environment or the biological limits of an organism. A simple example, focusing on pizza delivery drivers, is discussed below to illustrate these concepts.

Over the last two decades, the delivery of ready-to-eat pizzas to consumers' homes has become a booming business and has resulted in the creation of an army of pizza delivery drivers. A driver could have a number of goals in going about his or her work. Some obvious examples are to make absolutely as much money as possible (money), to make as much money as possible for the amount of time worked (money/time), or to make as much money as possible for the amount of time worked while minimizing wear and tear on personal vehicles, optimizing against potential future repair costs ([Money earned − repair and maintenance costs]/time). For this example, numerous decision variables could be modeled, such as time worked per week, shifts taken, routes driven, or driving velocity. Modeled benefits could include income, or status with managers, which could lead to improved working status or loss of work. Many costs are conceivable: for example, time costs, vehicle costs (gas and repairs), speeding tickets, or stress. Possible constraints include business hours, the maximum speed a vehicle can be driven, or the maximum number of hours a driver can work per week. The interplay of all of these factors could be used to model the optimal behavior of goal-oriented pizza delivery drivers.

Assume a driver has the goal of maximizing the total amount of money he or she makes in a week and must decide how many hours to work per week. Obviously, the more the driver works the more money he or she will make, so he or she will choose to work as many hours as possible given the operating constraints—business hours, other time commitments, or the number of other drivers employed. If we change the goal, however, to maximize work efficiency, that is, to make the most money possible for the amount of time worked, and instead allow the driver to choose the shifts to be worked, the solution may be different. If the driver is working solely for an hourly wage (a constraint assumption), then the shift taken should have no impact on the rate of income capture since it will be constant. In this case, a driver should be indifferent to the shifts he or she must work. If the driver is working for tips, however, the driver should choose the shifts that maximize the potential for earning tips, so the driver would likely choose shifts where the rate of pizza orders and delivery is maximized. Similarly, assume a driver is working for tips and is competing with other drivers working the same shifts. Assume the driver wants to maximize income per time worked and is deciding instead what speed to drive while delivering pizzas. In this case, the optimal solution may be to drive as fast as possible, given the constraints in place (e.g., the vehicle, speed limit, and/or police presence), to maximize the number of routes driven per unit time. If the driver has the same goal but is cognizant of potential repair costs, the driver may adjust speed to reduce the possibility of traffic accidents.

From this example, it should be clear how costs, benefits, goals, decision variables, and constraints interact in behavioral ecology to define optimal behaviors. Furthermore, this set of models demonstrates that there should not be a single optimal strategy for pizza delivery drivers. Behavioral strategies are expected to vary with respect to context. It should be noted that this example has optimality goals that would generally be perceived as "good" (e.g., maximize money), but optimal and good are two different things. Humans are certainly capable of making optimal decisions that may not be perceived as "good" in a moral or evolutionary sense. A pizza delivery driver might, for example, devise an optimal solution to getting fired. Different individuals may have different goals or operate under different constraints, and a single individual could have changing goals or constraints throughout their lifetime. A single model is unlikely to explain all of the variation in a particular behavior for all contexts.

## Technological Organization and Behavioral Ecology

Following Nelson (1991:57), technological organization is defined as "the selection and integration of strategies for making, using, transporting, and discarding tools and the materials needed for their manufacture and maintenance." The study of technological organization is, therefore, well suited to the use of formal models from behavioral ecology because decisions must be made at virtually every stage of stone tool production and use, and those decisions can be modeled as optimization problems. The concept of optimal technological strategies is not novel to the investigation of chipped stone technological organization; in fact, an entire book was devoted to the topic (Torrence 1989a). In the 1970s and early 1980s, Binford (1973, 1977, 1979) set the ball rolling by deriving principles that he believed were governing certain aspects of technological strategies used by hunter-gatherers in relation to environmental context. Inspired in part by ethnoarchaeological observations of the Nunamiut, Binford (1979:255) explored "the organizational alternatives within a technology which may be manipulated differently to effect acceptable adaptations to differing situations." Therefore, in the first studies of technological organization, the ties to behavioral ecology and optimality are evident. Although most traditional models of technological organization are fundamentally concerned with the optimization of some currency relative to some constraint or constraints, optimality is not always an explicit concern (Nelson 1991).

Models of technological organization have considered many aspects of lithic technology from procurement to discard, often emphasizing aspects of the design of tools and toolkits. Many constraints are evident in these discussions, but most commonplace are those imposed by mobility (Binford 1977, 1979; Kuhn 1989, 1995; Shott 1986), raw material availability (Andrefsky 1994; Bamforth 1986), food resource structure (Bleed 1986; Torrence 1983, 1989c), and social context (Wiessner 1983). Currencies are typically energy or utility, time (Binford 1977; Torrence 1983), risk (Bamforth and Bleed 1997; Torrence 1989c), or some combination of these. With only a very few exceptions (Brantingham and Kuhn 2001; Bright et al. 2002; Kuhn 1994; Metcalfe and Barlow 1992; Ugan et al. 2003), these models are informal, wherein generalized lawlike propositions are derived by verbal argument and are tested against ethnographic and/or archaeological cases. Though this approach has allowed us to make great strides in understanding the ways in which hunter-gatherers organize technology, it is problematic because the models tend to be so generalized that it often remains unclear whether the stated predictions follow directly from the implied goals, currencies, and constraints. Furthermore, because goals, currencies, and constraints are not formally stated, false contradictions can arise between models. An example is discussed below.

"Curation," as a property of technological systems, has received considerable attention from archaeologists, and the constraints shaping curated technologies have been a recurrent matter of debate (Bamforth 1986; Binford 1973, 1977, 1979, 1980; Hayden et al. 1996; Nash 1996; Shott 1996). When Binford (1973:242) introduced the concept, he defined curation with respect to "a tool once produced or purchased is carefully curated and transported to and from locations in direct relationship to the anticipated performance of different activities." In a later article (Binford 1977), he related curation to a form of efficiency (use-life per manufacture time) and noted that curation goes side by side with maintenance and recycling to enhance the tool efficiency by prolonging use-life. Curation was then contrasted with "expedient" or "non-curated" assemblages, which are characterized by the "manufacture, use, and abandonment of instrumental items in the immediate context of use" (Binford 1977:34). Binford (1977:35) went on to suggest that the degree to which tools are curated should relate to the organization of mobility, with logistically organized hunter-gatherers tending to depend more on curated technologies than those characterized by moving consumers to food, what later became known as the "forager" strategy (Binford 1980). The basic argument was that a greater reliance on curation was an optimizing solution to the problem of moving food to consumers because it increased the efficiency of tools (by prolonging use-life) in terms of the work they performed relative to the investment made in their manufacture.

In this model, the goal, decision variable, currency, and constraints are implicit but can be extracted. According to Binford, hunter-gatherers are attempting to maximize tool or toolkit efficiency (the goal). Efficiency, therefore, is the currency and is defined as the use-life of the tool, divided by the initial labor investment in time. It is unclear if maintenance and recycling costs come into play in calculating efficiency. The decision variable is difficult to define but obviously pertains to curation, which raises the sticky issue of what is meant by curation. Many authors have noted that curation encompasses many aspects of lithic technology (Bamforth 1986; Nash 1996; Nelson 1991; Shott 1996), but for now, we can define the decision variable as one that varies from curation to expediency, without specifying exactly what those terms mean. Finally, the decision is made relative to constraints imposed by the organization of mobility. Now the more difficult question: does the prediction that curated technologies should be associated with logistically organized hunter-gatherers hold? From the way the model is constructed, this question is difficult to answer. Assumptions in informal models are generally implicit, which may lead to a superficial appearance of a reasonable argument, but upon closer inspection, it can sometimes be shown that these assumptions are flawed (e.g., Surovell 2003b; Winterhalder and Goland 1993).

This problem highlights the difficulties encountered when informal narrative models are the epistemological norm. Obviously, the prediction of the model could be tested against an ethnographic or archaeological case, but even if the hypothesis is supported, we run the risk of being right for the wrong reason since we have no reason to believe our theoretical model is valid. The use of formal models alleviates only one of these problems. That is, we still risk the possibility of being right for the wrong reason, because of the possibility of equifinality, but at least we can rest assured that our predictions do derive from our theoretical construct. Returning to the problem of defining curation and expediency, from Binford's (1973, 1977) discussion of the terms it is impossible to decipher their exact meaning. Does curation equate to use-life, transport, maintenance, recycling, manufacture in anticipation of use, or the location of manufacture, use, and discard, or is curation some combination of all of these things (Shott 1996)? Does it apply to individual, tools, entire toolkits, or assemblages (Nash 1996)? If this model had been constructed formally, there might be no question as to the definition of curation, but in its original form the lack of specificity as to meaning of the decision variable subsequently led to considerable confusion over its use (Bamforth 1986; Nash 1996; Nelson 1991; Shott 1996). Almost thirty years after its inception, there is no consensus as to its meaning.

Contrasting Binford's original works with a later paper by Bamforth (1986) highlights a further problem with the use of informal models. Bamforth (1986:39) found Binford's curation concept somewhat unwieldy since it combined so many aspects of technology, and he argues reasonably that all of these behaviors may not act together to achieve the same goal. He further objected to the failure to consider the availability of raw materials for tool manufacture as a factor conditioning curation. Bamforth explored the effects of the organization of mobility and raw material availability on maintenance and recycling, two aspects of curation. Bamforth hypothesized that foragers are attempting to maximize efficiency, defined as fulfilling the "requirements of an activity or set of activities that constrain variation in all aspects of tool manufacture and use . . . with minimum effort" (Bamforth 1986:39). He argued that raw material availability is ultimately conditioning maintenance and recycling, and furthermore that raw material availability is a product both of the natural distribution of lithic raw materials and situational patterns of human behavior that "restrict access to raw material in specific contexts" (Bamforth 1986:40) such as settlement organization. Specifically, the argument was that the frequency of tool maintenance and recycling should increase in contexts of raw material scarcity since the cost of replacing tools is enhanced (Bamforth 1986:40).

Bamforth then tested this model against both ethnographic and archaeological evidence and found that raw material availability, as conditioned by settlement organization, is a good predictor of maintenance and recycling. Furthermore, high rates of maintenance and recycling in some cases are associated with settlement systems tending toward the "foraging" end of the spectrum, more so than for "collectors."

Although Bamforth's model suffers from problems similar to those of Binford's—poor definition of constraints and currencies, and unsubstantiated causal relationships—a second problem is highlighted when the two models are juxtaposed. Do Bamforth's model and findings invalidate Binford's? In one respect, they do, since it was found that maintenance and recycling, two aspects of curation, do not always correlate with the degree to which mobility is logistically organized. However, since the problems are defined differently, with respect to goals, currencies, and constraints, it could be argued that there is no contradiction whatsoever. After all, Binford's and Bamforth's "efficiency" are constructed differently, despite both models aiming to describe strategies that maximize efficiency. This is not to say that Binford's and Bamforth's models are necessarily incorrect or logically flawed, but that they are ultimately exploring slightly different things. Therefore, when informal models are used to explore the same or similar phenomena, contradictions that arise may simply be differences in semantics and may not be an invalidation or falsification of the original theoretical construct.

## Formal Models of Lithic Technology: A Short Review

To suggest that the use of formal models to explore technological strategies is novel or unique to this study would be disingenuous. To my knowledge, at least three such models have been proposed to date (Brantingham and Kuhn 2001; Kuhn 1994; Metcalfe and Barlow 1992; see also Bright et al. 2002; Ugan et al. 2003), and they are wide in scope, exploring very different aspects of lithic technology. Summaries of these models are presented here to emphasize the utility of formal models in the study of technological organization.

Metcalfe and Barlow (1992) are concerned with the use of resources characterized by two portions, one with utility and one lacking utility.

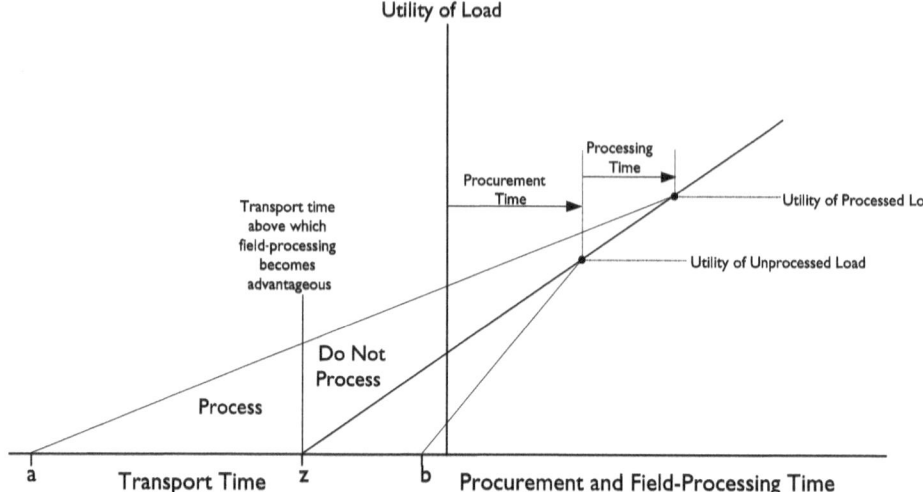

FIGURE 1.2. Graphical representation of the Metcalfe and Barlow (1992) field processing model. See text for description. Adapted from Metcalfe and Barlow (1992:fig. 1).

Nuts, with an inedible shell and an edible nutmeat, are a good example. In particular, they examine the trade-off for central place foragers in field processing and transport; that is, is it in a forager's interest to remove the no-utility portion in the field and transport a processed load of a resource to a central place, or alternatively should he or she transport a load of the resource unprocessed? They suggest this model could be applied to the use of lithic raw materials, since cobbles are often composed of low-utility cortical material and high-utility interior stone.

In the model, a forager attempts to maximize utility of a load per time spent foraging. Foraging time comprises transport time, procurement time, and field processing time. The decision variable is processing time, and the decision is constrained by the nature of the resource (the percentage of the load weight composed of the usable portion and the time needed to field process a load), and transport time, essentially how long it takes to make the round-trip from a central place to the foraging patch. The simplest form of the model predicts that for a given transport distance, the resource will either always be completely processed or not processed at all, a prediction similar to the zero-one rule of the classic diet breadth model (Charnov 1976a; Stephens and Krebs 1986:20–21).[1] A graphical representation of the basic model is presented in figure 1.2.

Transport time is on the left side of the x-axis, and procurement and field-processing time are on the right axis. The y-axis is the utility of the load. In this format, it is fairly simple to derive the optimal solution for a given resource and transport distance. First, the utility of a processed and unprocessed load are plotted with their respective procurement and processing times. For an unprocessed load, only procurement time comes into play, but for a processed load, both procurement and processing time will combine to prolong foraging time. Next, the transport time is plotted on the left side of the x-axis, and lines are drawn from the point on the x-axis to the points describing the processed and unprocessed loads. The line with the greater slope is the optimizing solution since slope of the line (utility/time) represents the currency. The transport distance above which field processing becomes optimal is shown as point z in figure 1.2, and it is fairly simple to demonstrate that it is the x-intercept of the line connecting the relative utility of the processed and unprocessed loads. The model predicts that a given load should remain unprocessed for short transport times (point b, fig. 1.2) and should be processed for longer transport times (point a, fig. 1.2).

Using a slightly more complex version of the same model, Metcalfe and Barlow (1992:352) develop expectations for field processing of lithic raw materials for transport to a central location, such as a residential camp. Essentially, the model predicts that as the distance to a raw material quarry area increases, the degree to which raw materials are "field processed," or trimmed of their low-utility portions, should increase. Furthermore, they suggest a test involving the study of archaeological assemblages at increasing distances from a quarry area, or research aimed at looking at staging in artifacts of different raw materials within a single site (e.g., Beck et al. 2002).

While Metcalfe and Barlow were concerned with behavior toward the procurement end of artifact production and use, Kuhn (1994) developed a formal model exploring two technological trade-offs in the design of mobile toolkits. In particular, Kuhn asks, "Presupposing that mobile toolkits are designed to maximize durability and functional versatility, while simultaneously minimizing weight (1) should people carry cores or tools/tool blanks, and (2) should they transport a few large artifacts or a large number of small ones?" (Kuhn 1994:426). In Kuhn's model, technological options in the design of mobile toolkits are seen as trade-offs, where alternative strategies have associated costs and benefits that can be modeled with respect to a currency to define optimizing behaviors.

Kuhn's (1994) model assumes a currency of utility divided by mass, and thus the goal is to find the technological solution that maximizes this quantity for a toolkit. Utility is defined as the potential to produce usable flake edges and is measured relative to a minimum usable size for tools and cores. Because of the morphological properties of flake tools and cores, utility for tools is calculated one- or two-dimensionally as a function of flake length or area, while utility for cores is calculated three-dimensionally and is deducted by a coefficient representing the percentage of raw material that will go unutilized (e.g., preparation and waste flakes). Transport costs for both tools and cores, however, are calculated in proportion to volume.

From this model, Kuhn (1994) derives two predictions. First, mobile toolkits should comprise tools or tool blanks rather than cores because they have more usable flake edge per unit mass. This observation derives from the reasonable assumption that core reduction, no matter how efficient, cannot perfectly convert tool stone into usable tool blanks without producing waste. Therefore, the transport of cores necessarily entails the transport of a greater proportion of unusable raw material than the transport of flake tools. The second prediction is that transported tool blanks should optimally be between 1.5 and 3.0 times their minimum usable size, dependent on a one-dimensional or two-dimensional measure of utility, respectively (fig. 1.3).[2] In chapter 5, I return to this model and provide archaeological tests of its predictions.

Although a number of papers tout the benefits of various forms of lithic reduction (e.g., Ahler and Geib 2000; Clarke 1976; Kelly 1988; Sheets and Muto 1972), relatively few studies consider the costs associated with them (e.g., Parry 1994; Parry and Kelly 1987; Rasic and Andrefsky 2001). Ahler and Geib (2000), for example, argue that the fluting of Folsom points extended the use-life of projectile tips by allowing the user to minimize tip breakage through a specialized hafting technique. While this model is an interesting explanation of Folsom fluting, it leaves one wondering why fluting techniques were ever abandoned since there are apparently great benefits, but no significant costs, to this form of basal thinning. In contrast to these studies, Brantingham and Kuhn (2001) explore optimization trade-offs within a single reduction strategy,

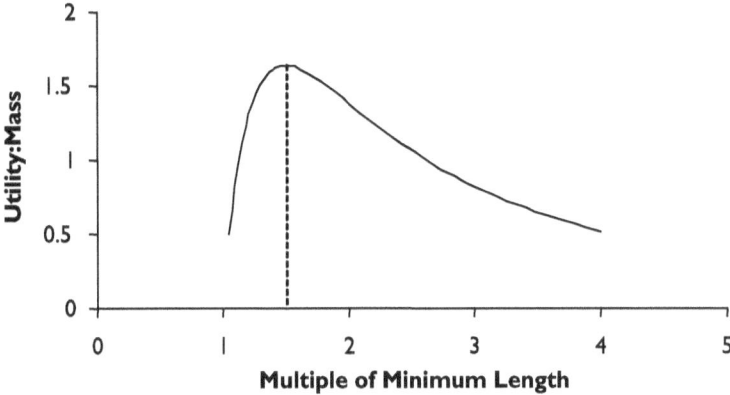

FIGURE 1.3. Graphical representation of the Kuhn (1994) model of tool blank size vs. utility:mass, where utility is measured one-dimensionally. Flake size is expressed as a multiple of the minimum usable size of a tool. The model predicts an optimal tool blank size of 1.5 times the minimum usable portion (dashed line). Adapted from Kuhn (1994:fig. 5).

namely Levallois technology, using a formal mathematical model of core reduction.

Instructional flintknapping books often try to illustrate the proper, or optimal, striking angles and platform morphologies for removal of certain types of flakes (e.g., Whittaker 1994), shedding light on issues of optimality that pervade even the most fundamental behavior in chipped stone technology. Brantingham and Kuhn (2001) use a formal model to explore similar options in Levallois reduction but instead derive solutions for maximizing or minimizing currencies that may have guided the use of this technology. A graphical representation of the model is shown in figure 1.4. The authors begin with a two-dimensional elliptical representation of a well-rounded cobble to be modeled as core. Then, based on two starting cobble shapes, they derive the optimal platform position and angles relative to three goals and currencies: minimization of wasted raw material, maximization of the number of usable products, and maximization of product edge length. They find that the strategies that best mimic Levallois technology (steep platform angles and platforms situated near the edge of a cobble) simultaneously maximize the amount of usable edge produced and minimize raw material waste. They also note, however, that this reduction strategy does not optimize the number of

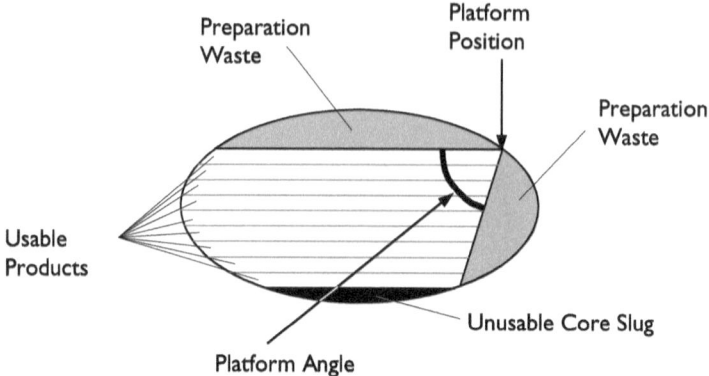

FIGURE 1.4. Graphical representation of Brantingham and Kuhn's (2001) formal model of Levallois reduction. Areas in gray and black represent wasted raw material. Platform position and angle are decision variables. Currencies are wasted raw material, number of usable products, and total edge length of usable products. Adapted from Brantingham and Kuhn (2001:figs. 3 and 5).

usable products, hinting at some of the trade-offs inherent to flintknapping decisions. Although they did not test the predictions of the models, one could envision a simple test comparing observed platform angles of a large sample of Levallois cores to their derived optimal solutions relative to their various currencies.

From these examples, three general observations can be made of formal models of lithic technology. First, the models explore three very different aspects of technological organization, from procurement of raw materials to decisions made in reduction to the design of toolkits for mobile foragers. This highlights the versatility of formal modeling and shows that its potential to explore aspects of technology is limited only by our imagination. Second, because the models are very explicit in stating assumptions, currencies, constraints, and goals, there is no ambiguity in definition, and their predictions are explicit and logically valid, assuming only that no errors were made in the formulation of the models. Lastly, it is interesting that none of the three models was actually tested with archaeological data, perhaps hinting at a downside to formal approaches. Could their assumptions and requirements be so strict as to limit or even prohibit testing with archaeological data? This question will be explored in the following section.

## The Costs of a Mathematical View of the World?

> It should then be possible to devise an "optimal tool-kit" in much the same way that previous scholars have modeled subsistence using principles from optimal foraging theory in ecology. . . . At this preliminary stage in the development of theory for technology, however, such detailed analyses are not really necessary or appropriate, although the formation of more specific models should be the ultimate goal. (Torrence 1983:14)

Although I agree with the final statement that, ultimately, we should be developing a formal theory of lithic technology, I am left wondering why, in the opinion of Torrence, such models are not "necessary or appropriate," at least at that point in theoretical time. In this section, I will discuss some of the problems associated with formal theory but ultimately argue that these problems are not novel to mathematical models but are difficulties archaeologists have faced all along.

Beginning with the observation that none of the three models presented in the previous section was published with an accompanying archaeological test, we can ask, why not? Metcalfe and Barlow (1992:352) provide one hint with reference to a possible test of their field-processing model: "The emphasis is on explaining variation; using the implications of the model to interpret some invariant aspect of an assemblage (the presence or absence of a particular resource) is not likely to be profitable. To do so with any rigor would require estimating the various parameters of the model with a level of precision unlikely ever to be available from the archaeological record" (Metcalfe and Barlow 1992:352). The archaeological record is very coarse-grained and does not easily yield crisp quantitative estimates of constraints, costs, and benefits as required by rigorous models. Take, for example, a model that attempts to explain the optimal number of flake tools to be carried given a trade-off between attempting to maximize potential energy capture and to minimize the mass of tool stone to be carried. Estimating the mass of a set of flake tools of a given size should not be difficult, but how much energy is an endscraper 30 mm in length capable of capturing? Obviously, endscrapers are not used directly to capture energy. Instead they are used to process hides, which are made into clothing, shelter, and tools, without which people in most environments would not be able to capture much energy. However, endscraper length should be proportional to its potential to capture energy because longer endscrapers should have longer use-lives and therefore can be used to process greater areas of hide. Thus, the length of endscrapers could be used as a proxy variable for "energy capture potential." This issue emphasizes an important fact: the testing of most formal models of lithic technology must often rely on proxy measures of currencies and constraints. Therefore, it could be argued that when we begin to compound all of these uncertainties, a formal mathematical model loses its formality in the testing phase because it cannot be demonstrated that its assumptions have been met, or that its relevant variables have been adequately quantified.

There is some truth to this statement, but I would argue that the utility of formal models is not that they can be flawlessly tested and applied archaeologically, but as discussed above, that they are logically valid. Although we may never be able to measure the potential of a déjeté scraper to capture energy, or the energetic cost of an Elko point, we can know with certainty how such properties of artifacts should relate to adaptation in specific environmental contexts under specific optimization goals. The nature of the archaeological record will require proxy measures of key variables, but this insurmountable truth in no way should influence the rigor with which we develop and apply formal models. Furthermore, it should be apparent that this problem is not unique to formal models. They pervade all models of technological organization. For example, Torrence (1983) argues that tools should be designed with increasing complexity as hunter-gatherers become increasingly time-stressed. Obviously, the constraint of time-stress in Torrence's informal model is one that would be nearly impossible to directly measure archaeologically, and to measure even ethnographically would require large amounts of data that are often unavailable. In lieu of this, Torrence uses latitude as a proxy measure, applying a handful of linking arguments to justify its use.

A more pressing issue relates to the simplicity of the assumptions that form the basis of formal models, an important issue that can be examined in light of an old joke about the nature of scientific research:

> A group of dairy farmers pools their money and hires a group of scientists to derive the optimal diet to maximize milk production

for their heifers. For six months, the scientists research and refine a model of diet, nutrition, and milk production and return to the anxious farmers with their results. They begin their presentation with the statement, "Assume a spherical cow."[3]

The real irony of this joke is that since the shape of a cow should have little to do with its ability to convert food into milk, the simplifying assumption of round cows may be justified. If the scientists instead were studying bovine thermoregulation, then the assumption would be highly problematic since the maintenance of body temperature is in part a product of the ratio of surface area to volume, and spherical cows and real-world cows would differ dramatically in this regard.

Although simplifying assumptions are often a necessary component of formal modeling, it is important to consider under what conditions they may seriously bias our predictions of the archaeological record. The "spherical cow problem" is one that rears its curvy head from time to time in this study, but again, it is not unique to formal models. The transparent nature of formal models often allows us to see the assumed "spherical cow" in clear daylight, whereas it often remains camouflaged between the lines of informal models. Regarding the simplicity of models in behavioral ecology, Winterhalder and Smith perhaps put it best: "Models, however, may invite criticism because of their simplicity. Superficially there is an intuitive appeal to such critiques. The models of evolutionary ecologists are orders of magnitude short of the complexity and variety of phenomena they purport to explain. They invite the accusation of simplemindedness and reductionism. But simple is not simple-minded. Simple models are a necessary, not a temporary or primitive stage of scientific development" (Winterhalder and Smith 1992:13–14).

## Organization of the Study

In chapter 2 I provide paleoenvironmental and cultural historical background to the study. I define the spatial and temporal ranges of Folsom and Goshen and identify trends in subsistence, mobility, and technology. I also provide general descriptions of the sites and assemblages that form the centerpiece of the study. In chapters 3 and 4, I develop assemblage-scale measures of occupation span that are later used to monitor technological variability. I also develop methods for differentiating multiple occupations from single occupations for archaeological components. The remainder of the study explores various issues surrounding the use of stone tools ordered in technological progression from lithic procurement to discard. In particular, in chapter 5 I focus on raw material supply, developing and testing a model of lithic surplus accumulation. In chapter 6 I explore the optimal design of flake tools, cores, and bifaces as constituents of mobile toolkits. Finally, in chapter 7, I model the discard of debitage as the by-product of factors governing tool and biface production and reduction.

# 2

# Late Pleistocene Foragers of the Northern Plains and Rocky Mountains

IN THIS CHAPTER I provide a cultural historical background to the study. The artifact assemblages examined for this study include 48,501 pieces from five late Pleistocene archaeological sites in Wyoming and Colorado: Agate Basin (Area 2, Folsom component), Carter/Kerr-McGee (Folsom component), Krmpotich, Barger Gulch (Locality B), and Upper Twin Mountain. In the artifact sample, 99.8 percent of the artifacts are from Folsom sites, and only 87 artifacts, or 0.2 percent, come from the Upper Twin Mountain Goshen bison kill. Therefore, the vast majority of the data used in this study are from Folsom residential occupations, or campsites. The Upper Twin Mountain site, although considered typologically Goshen, was included in the study sample to incorporate lithic data from a comparably aged Paleoindian bison kill because research constraints did not allow the analysis of a Folsom kill site assemblage. Inclusion of a kill assemblage permits exploration of variability in lithic technology relating to site function.

Why these particular assemblages? The models I develop in this study are extremely generalized, and as such, they could in theory be applied to any situation in which stone tools were made and used, from the Pliocene to the present. To test these models requires only a set of lithic assemblages derived from occupations that varied in length, since my primary concern is exploring the effects of variation in residential mobility (and thus occupation span) on lithic technology. As will be shown in the following two chapters, the lithic assemblages analyzed for this study do meet this requirement, so they are well suited to the job. It is my hope that future research will further examine and test the models and hypotheses developed in this study using data collected from other contexts.

In some ways, however, they are not ideal. Foremost, the study sample includes only five sites, only four of which are campsites. Small samples severely limit one's ability to detect statistically significant trends, and I rely on assemblage-scale measures of technological variability. Finding significant correlations with only four or five data points requires extremely strong patterning. The sample used in this study is one of the largest multisite comparative analyses of Paleoindian artifacts ever undertaken. A related drawback of the study sample is that the sample size for two sites (Carter/Kerr-McGee and Upper Twin Mountain) is limited. The entire analyzed assemblage from Upper Twin Mountain includes only 87 artifacts; the Carter/Kerr-McGee assemblages include 1,990 pieces. Because of constraints imposed by variation in excavation methods and more so by methods used in my analysis of these assemblages (see Surovell 2003a: appendix I), I often rely on a reduced sample using only artifacts greater than or equal to 20 mm in length. The reduced sample from Upper Twin Mountain includes only 12 artifacts and the Carter/Kerr McGee sample only 40 artifacts. Of course, both of these sites represent relatively short occupations, and in short occupations limited numbers of artifacts are discarded. Therefore, while these small sample sizes can definitely be viewed as a limitation to the study, they also reflect the nature of the archaeological record. To overcome the limitations of the study sample, when possible I also include published data from other sources. I now turn to the cultural history of Folsom and Goshen.

## Folsom and Goshen in Space and Time

In this section I explore the spatial and temporal bounds of the Folsom and Goshen complexes. The intent is to demonstrate that the analytical units of "Folsom" and "Goshen" overlap considerably in these two dimensions. In fact, Goshen and Folsom overlap considerably in space-time, providing a unifying framework for the sites incorporated into this study. It is not my intent to address what the terms "Folsom" and "Goshen" mean in a sociocultural sense other than to note they represent two projectile point styles characterized by shared morphological attributes. I do not use the terms to necessarily represent one common, two similar, or two distinct "cultures" in the sense of shared values, language, subsistence, material goods, and so on. I refer the reader to LeTourneau (1998) for a discussion of the history, goals, and problems associated with Folsom and Paleoindian projectile point typology (see also Meltzer 2006a).

Folsom projectile points are perhaps one of the most recognizable artifact forms of the prehistoric record, usually characterized by two longitudinal channel flakes removed from the base, which extending almost the entire length give the point a "fluted" appearance. Folsom points typically exhibit parallel sides or sides that expand slightly from the base, basal lateral edge grinding, fine marginal pressure flaking, and a concave base with well-defined ears (fig. 2.1). The qualifying words of "usually" and "typically" are necessary because there is considerable variability within Folsom points with respect to size, outline, and flaking patterns (Hofman 1992; Judge 1973). Even fluting, the most distinctive characteristic of Folsom points, shows many variants, including points fluted on one side, unfluted points, and pseudo-fluted points (see fig. 2.12b), in which the ventral surface from the original flake is retained to superficially give the point a fluted appearance (Frison and Stanford 1982a; Hofman 1992; Wilmsen and Roberts 1984).

Although the vast majority of excavated Folsom sites occur in the western Great Plains and Rocky Mountains of the United States from Wyoming to Texas, Folsom projectile points are known to span a significantly larger area (fig. 2.1). The westernmost Folsom finds occur in northeastern Arizona (Faught and Freeman 1998; Wendorf and Thomas 1951), Utah (Gunnerson 1956; Hunt and Tanner 1960; Schroedl 2000), and Idaho (Miller 1982). Folsom sites are known as far north as Montana (Davis and Greiser 1992; Forbis and Sperry 1952), southern Alberta (Bliss 1939; Gryba 1985; Kooyman et al. 2001), and Saskatchewan (Ebell 1970; Howard 1939; Kehoe 1966). In the east, Folsom stretches into Wisconsin, Iowa, Illinois, and possibly even far western Indiana (Billeck 1998; Morrow and Morrow 1999; Munson 1990). In the Southern Plains, it is known to occur from far eastern Oklahoma and Texas (Blackmar 2001; Hofman 1990) to the Mexican states of Tamaulipas, Nuevo Leon, and Chihuahua, but Folsom points are rare in Mexico south of the Rio Grande (Aveleyra 1961; Epstein 1961, 1969:xiii; Sanchez 2001). Therefore, the Folsom complex (defined by the presence of Folsom projectile points) is known from three modern countries, spanning approximately 30 degrees of latitude and 25 degrees of longitude. Environmentally, it stretches from the Gulf coastal plain of eastern Texas to the Northern Plains of southern Canada, and from the Snake River Plain, eastern Great Basin, Rocky Mountains, Colorado Plateau, and Basin and Range in the west to the tall grass prairies,

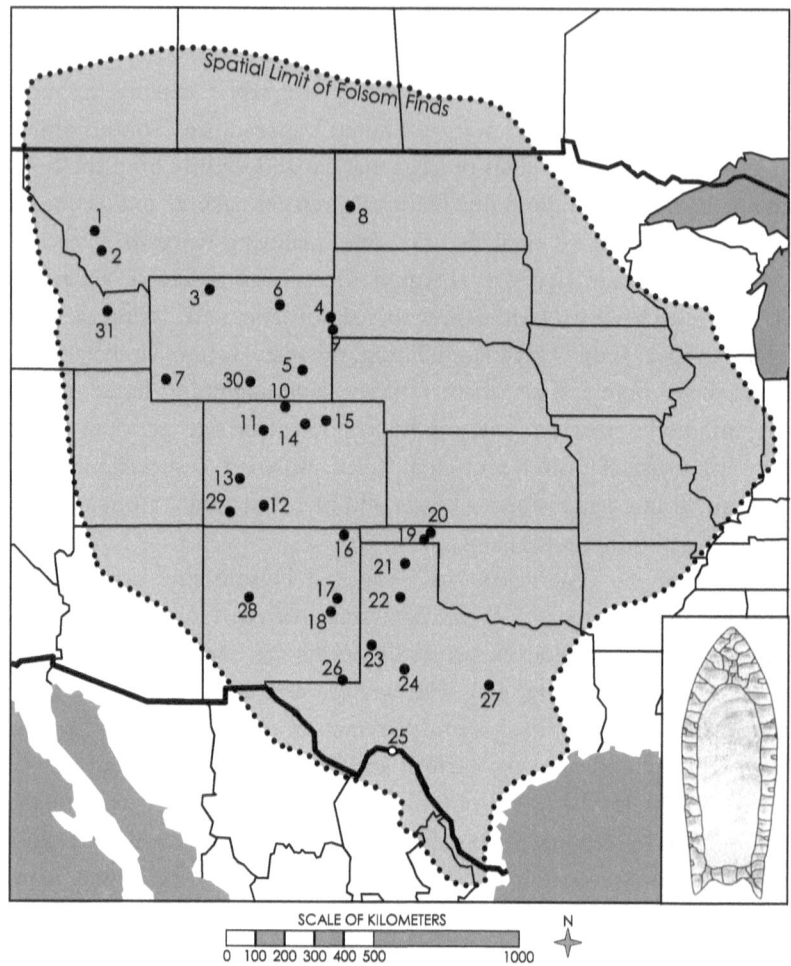

FIGURE 2.1. The total geographic area encompassed by Folsom point finds (shaded gray) with a sample of Folsom site locations shown as dots: 1. MacHaffie; 2. Indian Creek; 3. Hanson; 4. Agate Basin; 5. Hell Gap; 6. Carter/Kerr-McGee; 7. Krmpotich; 8. Lake Ilo sites (Bobtail Wolf, Big Black, and Young-Man-Chief); 9. Jim Pitts; 10. Lindenmeier, Johnson; 11. Middle Park sites (e.g., Barger Gulch, Crying Woman, Lower Twin Mountain); 12. San Luis Valley sites (e.g., Stewart's Cattle Guard, Reddin, Zapata, Linger); 13. Mountaineer; 14. Powars; 15. Fowler-Parrish; 16. Folsom; 17. Blackwater Draw; 18. Elida; 19. Waugh; 20. Cooper; 21. Lipscomb; 22. Lake Theo; 23. Lubbock Lake; 24. Adair-Steadman; 25. Bonfire Shelter; 26. Shifting Sands; 27. Horn Shelter; 28. Rio Rancho; 29. Black Mountain; 30. Rattlesnake Pass; 31. Wasden. Inset shows a generalized drawing of a Folsom projectile point (adapted from Wormington 1957:fig. 10).

rolling plains, and woodlands in the east. In elevation, Folsom sites extend from near sea level elevations in southern Texas to over 3,000 m in the Colorado Rockies (Jodry et al. 1996). Although Folsom points are found in an area exceeding 3,000,000 km² with significant environmental variability, what likely tied together this vast area was the presence of *Bison antiquus* (Munson 1990). The range of *B. antiquus* fossil remains, however, far exceeds that of Folsom (McDonald 1981:fig. 20).

In contrast to Folsom, Goshen projectile points are found over a much smaller area, in large part because of the type definition. Goshen is a regional phenomenon, by definition limited to the Northern Plains and Rocky Mountains. The excavation of the Mill Iron site in southeastern Montana from 1984 to 1988 revived the concept of a Goshen-type projectile point (Frison 1991a; Frison et al. 1996), originally proposed based on projectile points recovered during excavations at the Hell Gap site in southeastern Wyoming (Irwin-Williams et al. 1973). Much confusion as to the temporal and typological placement of this type has existed since its initial proposal (Frison 1991a; Frison et al. 1996; Haynes 1991a, 1993; Sellet 1999). Projectile points morphologically indistinguishable from those recovered at Mill Iron were found forty years earlier at the Plainview site in the panhandle of Texas (Sellards et al. 1947), and even at that time Krieger went to great lengths to argue that they should be designated by their own type, Plainview, which is still widely recognized. Irwin (1971) and Haynes (1991a) suggest that we should abandon the name "Goshen" altogether and use "Plainview" instead, since the latter was first defined. Frison et al. (1996:206) prefer to retain the Goshen type since "radiocarbon and stratigraphic evidence from the Northern Plains strongly indicate a pre-Folsom age for Goshen, while Plainview on the Southern Plains is believed to be of post-Folsom age." The age ranges for Plainview on the Southern Plains and Goshen on the Northern Plains, however, are far from resolved (Haynes 1991a; Holliday 1997, 2000; Holliday et al. 1999; Sellet 1999, 2001). For the purposes of this study, I will retain the use of the term Goshen, as it is commonly used in the literature, to refer to projectile points sharing Goshen/Plainview morphology recovered from the northwestern Plains (Wyoming, Montana, Alberta, and the western Dakotas) and Colorado Rocky Mountains.

Goshen projectile points are typically lanceolate in form and slightly basally indented (fig. 2.2). They generally have parallel sides or are slightly

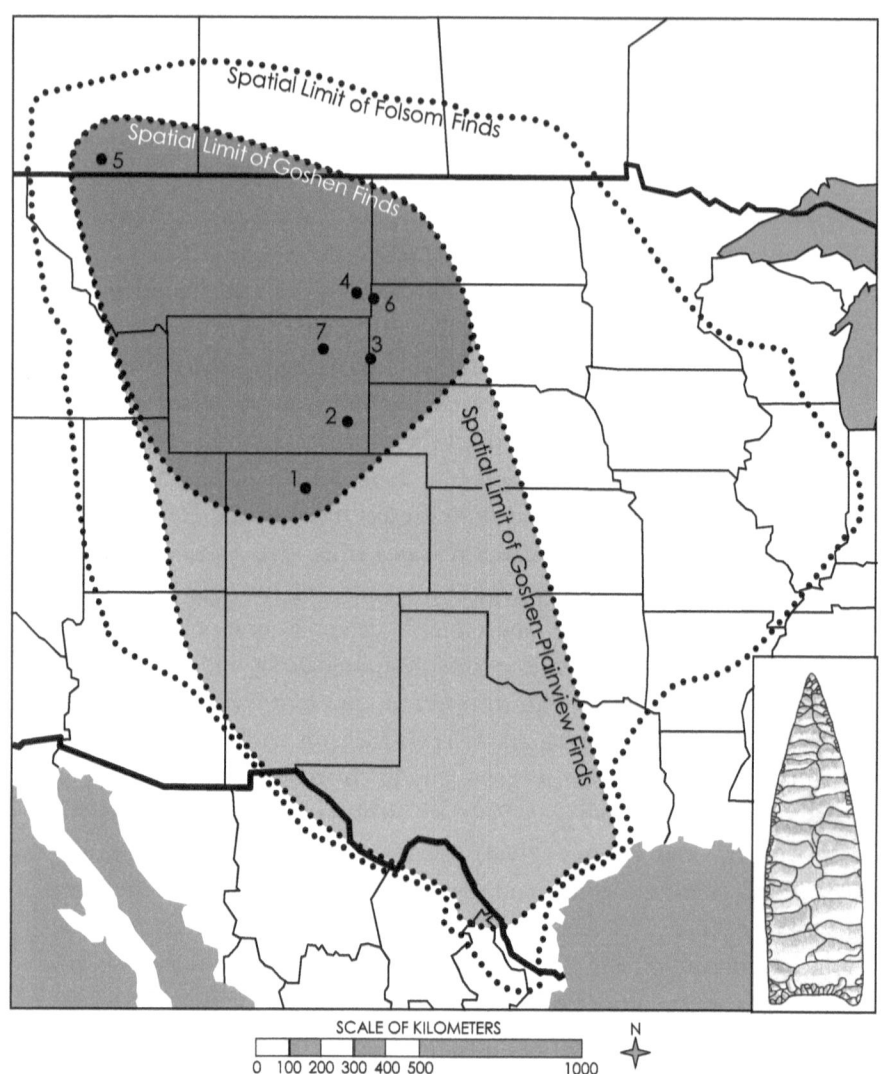

FIGURE 2.2. The total geographic area encompassed by Goshen projectile points (shaded dark gray) relative to Folsom (unshaded) point finds, with a sample of Goshen site locations shown as dots: 1. Middle Park sites (e.g., Upper Twin Mountain, Barger Gulch); 2. Hell Gap; 3. Jim Pitts; 4. Mill Iron; 5. Wally's Beach; 6. Ghost; 7. Carter/Kerr-McGee. Light gray area shows the rough distribution of Goshen finds inclusive of Plainview. Inset shows a generalized drawing of a Goshen projectile point (adapted from Bradley and Frison 1996:fig. 4.3d).

convex. In outline, they resemble both Clovis and Folsom points, but Bradley and Frison (1996) ally Goshen points more closely with Folsom points, as does Krieger (Sellards et al. 1947) for Plainviews. The blades are characterized by transverse or collateral pressure flaking with ground basal margins. Goshen points are not fluted but are commonly thinned basally with single or multiple pressure flakes. Basal ears are also apparent on some specimens.

Relatively few Goshen finds or sites have been reported, likely because of the relatively recent revival of the type. As discussed above, the distribution of Goshen, as shown in figure 2.2, is a function of its definition, rather than the geographic range of the Goshen-Plainview projectile point morphology. Goshen points have been recovered as far north as southern Alberta (Gryba 1985:fig. 3; Kooyman et al. 2001). At least thirty Goshen points, most of which derive from the Jim Pitts and Ghost sites, have been recovered from western South Dakota (Donohue and Sellet 2001; Fosha and Sellet 2000; Sellet and Fosha 2000). Goshen sites in Wyoming and Montana include Hell Gap (Irwin-Williams et al. 1973) and Carter/Kerr-McGee (Frison 1984). The southern boundary of Goshen is defined by a cluster of sites in Middle Park, Colorado, including Upper and Lower Twin Mountain, Barger Gulch (Localities A, D, and I), and Hay Springs (Kornfeld and Frison 2000; Kornfeld et al. 1999; Waguespack et al. 2006).

Goshen is entirely circumscribed in space by Folsom, being restricted to the northwestern portion of the Folsom range (fig. 2.2). Although the definition of Goshen (in distinction from Plainview) ensures this to be the outcome, the use of a definition of Goshen that is independent (i.e., inclusive of Plainview finds) would still result in the same pattern. In that case, Goshen would stretch as far south as northern Mexico (Epstein 1969) but would still not exceed the Folsom range. Put another way, projectile points sharing Goshen morphology have not been recovered in areas outside of that defined by Folsom. Having established that Goshen and Folsom occupy the same space, I now demonstrate the same finding holds for time.

Over one hundred $^{14}$C dates have been produced from Folsom contexts, as compared to at least twenty-one for Goshen (Donohue and Sellet 2001; Frison 1996b; Haynes 1991a; Holliday 2000; Kornfeld et al. 1999; Stanford 1999). Although the sample of dates from Folsom sites is quite large, the Folsom period is by no means well dated, at least in absolute time.

Consistent stratigraphic relationships have Folsom firmly dated relative to other Paleoindian components with the important exception of Goshen, but radiocarbon dates from Folsom sites are highly variable. For example, dates from the Lipscomb site range from 5,775 ± 110 BP (Beta 28193/ETH-4618) produced on charcoal humates to 13,440 ± 170 BP (NZA-1091) produced on charcoal (Hofman 1995). Similarly, dates on dispersed charcoal from Barger Gulch Locality B have ranged in age from 6,880 ± 60 (Beta-155404) to 10,770 ± 170 BP (Beta-173385) (Mayer et al. 2005; Surovell et al. 2005). These age ranges point to the many problems inherent in radiocarbon dating of archaeological sites of primary importance, problems of sample contamination and poor temporal association between the event dated and the event intended to be dated. These issues emphasize the need for careful selection of radiocarbon dates by material and context to arrive at an appropriate age range for any technological complex (e.g., Haynes 1991a; Taylor et al. 1996).

Haynes estimates the Folsom complex to range from approximately 10,950 to 10,250 BP (Haynes 1992, 1993; Haynes et al. 1992), and Holliday (1997) suggests a similar but slightly compressed age range for Folsom, 10,800–10,200 BP. Figure 2.3 presents a sample of site-averaged radiocarbon dates on charcoal or bone collagen from Folsom sites. These dates were taken from the total pool of dates for Folsom sites by elimination of outliers and dates from questionable contexts (Frison 1982c; Haynes et al. 1992; Hofman 1995; Holliday et al. 1983, 1985; Johnson 1987; Meltzer and Holliday 2006; Stiger 2006; Surovell et al. 2005; Wilmsen and Roberts 1984). Averages were calculated using the Long and Rippeteau (1974) method. This sample of dates provides an estimate of the Folsom period that corresponds closely with Haynes's estimate (fig. 2.3). Thus, the Folsom period correlates well with the Younger Dryas (ca. 11,000 to 10,000 $^{14}$C yr BP; 12,940 to 11,640 cal yr BP), a 1,300-year-long return to glacial conditions that affected the Northern Hemisphere (Broecker et al. 1988, 1989; Fiedel 1999; Gosse et al. 1995; Polyak et al. 2004; Reasoner and Jodry 2000). It is interesting to note, however, that the most precisely dated sites (Folsom, Mountaineer, and Barger Gulch, Locality B) tightly cluster in a fairly narrow time range from 10,400 to 10,600 $^{14}$C yr BP.

The initial indication of the age of the Goshen complex, derived from stratigraphic relationships observed in excavations at the Hell Gap site, suggested that Goshen predated Folsom (Irwin-Williams et al. 1973).

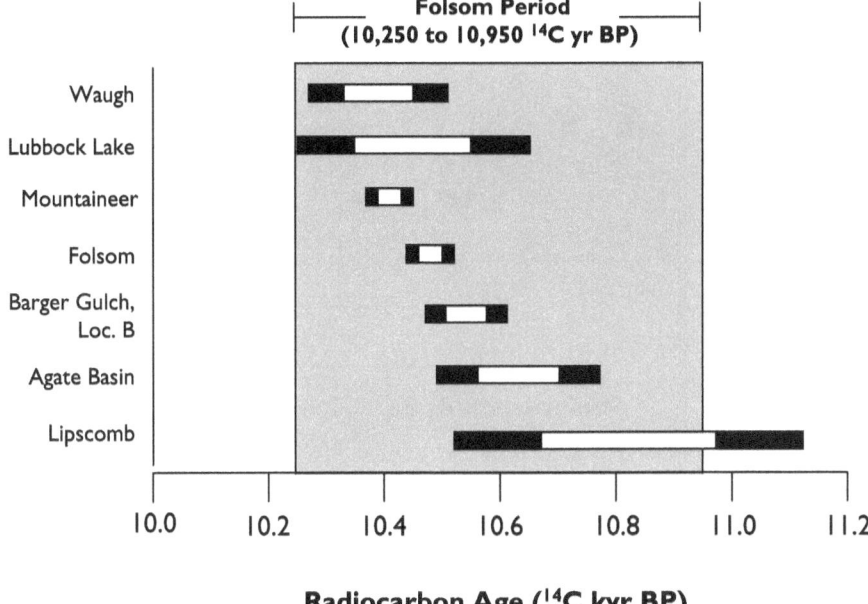

FIGURE 2.3. Site-averaged charcoal or bone collagen radiocarbon dates for seven Folsom sites. White and black boxes show the 1σ and 2σ age ranges, respectively. Gray box shows estimated time range for Folsom from Haynes (Haynes 1992, 1993; Haynes et al. 1992).

Further stratigraphic support for this idea came from Carter/Kerr-McGee (Frison 1984; Frison et al. 1996). Although Frison (1984) initially referred to the Carter/Kerr-McGee points, found approximately 10 cm beneath the Folsom level, as Clovis, he later suggested that one of the points might in fact be Goshen (Frison et al. 1996). Radiocarbon dates from Upper Twin Mountain suggested a Folsom age (Kornfeld et al. 1999) at approximately $^{14}$C yr 10,300 BP, while dates on dispersed charcoal from the Jim Pitts site in southwestern South Dakota produced a post-Folsom age of 10,150 BP. In his reanalysis of artifacts and cultural stratigraphy from Locality 1 of Hell Gap, Sellet (1999:115–119, 2001) found that the stratigraphic separation of Folsom and Goshen was not as clear as had originally been reported. Therefore, the chronological position of Goshen relative to Folsom remains very much unresolved.

Figure 2.4 illustrates a sample of radiocarbon dates available from Goshen sites (dates from Donohue and Sellet 2001; Frison 1996b; Haynes

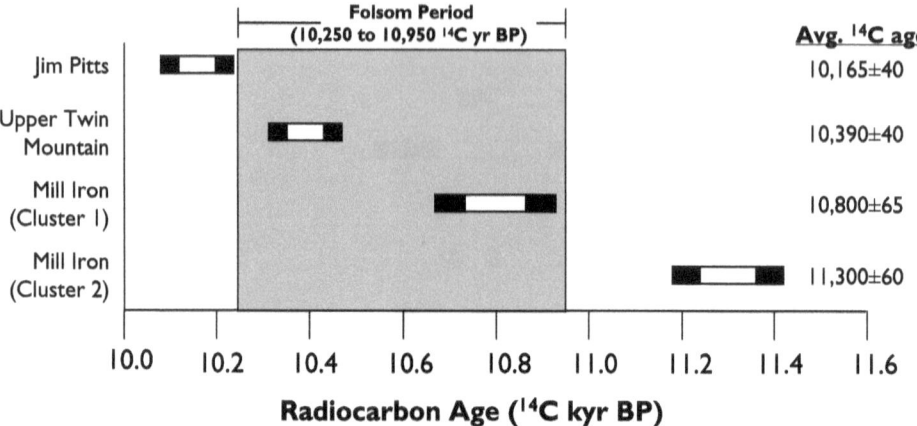

FIGURE 2.4. Site-averaged charcoal radiocarbon dates for three Goshen sites. White and black boxes show the 1σ and 2σ age ranges, respectively. Gray box shows the estimated time range for Folsom from Haynes (Haynes 1992, 1993; Haynes et al. 1992). Cluster 1 and Cluster 2 for the Mill Iron site refer to clusters in radiocarbon ages (not spatial clusters).

1991a; Kornfeld et al. 1999). As I have done for Folsom above, site averages were calculated after suspect dates were removed. These include an anomalous 23,720 ± 220 BP (AA-3668) charcoal date from Mill Iron that suffers from possible lignite contamination (Haynes 1991a), and a bone date from Upper Twin Mountain (8,090 ± 60 BP, Beta-76593) that did not undergo rigorous pretreatment (Kornfeld et al. 1999:658). Because the dates from Mill Iron form two discrete clusters, two averages were calculated.

From figure 2.4, it appears that Goshen entirely encompasses Folsom in time, but this conclusion may be unwarranted. Although there exists the potential for multiple overlapping occupations at Mill Iron (Haynes 1991a), only a single component was detected (Frison 1996b; Frison et al. 1996). Given that the site produced an anomalously old date and the potential for old wood and lignite contamination (Frison 1996b; Haynes 1991a), perhaps the younger cluster at 10,800 ± 65 provides a better age estimate for the occupation. Although six dates from the Jim Pitts site do form a relatively tight cluster, Donohue and Sellet (2001:20) report that sixteen samples were dated, and "of these, six non-intrusive samples are directly associated with the bone bed." Therefore, there appears to be little independent evidence to suggest that the charcoal samples that exceed 10,000 BP in age are associated with the Goshen occupation, something Donohue and Sellet (2001:20) readily admit. Therefore, the dates from Upper Twin Mountain are perhaps the most reliable age estimate for Goshen, despite being bone dates. Notably, these samples underwent the rigorous pretreatment protocol developed by Stafford et al. (1991) and fall squarely within the Folsom age range. With these considerations, Folsom and Goshen in the northwestern Plains and Rocky Mountains appear to overlap closely in both space and time. For the purposes of this study, then, I treat Folsom and Goshen as a single analytical unit, but by doing so I do not mean to imply that they could not be treated otherwise for other purposes.

## Folsom and Goshen Lifeways

The following reviews of Folsom and Goshen subsistence, mobility, and technology rely heavily on data from Folsom sites for the reason that significantly more Folsom sites have been excavated. The extent that generalizations made of Folsom also apply to Goshen is unknown, but again, this is not significant since it is time and space that are held constant in this study, not projectile point morphology.

### *Subsistence*

If a single word can be used to sum up Folsom and Goshen subsistence, it is "bison." Bones of *Bison antiquus* are present in virtually every Folsom and Goshen site containing well-preserved faunal remains that has been excavated to date (Amick 1994a; MacDonald 1998a), and numerous authors have argued that many aspects of Folsom lifeways, in particular, can be explained by focal predation upon this species. For example, Hofman (1999a:384) states, "Folsom culture, society, economy, land use patterns, and technology would have been radically different had bison been eliminated from the repertoire of prey species." Likewise, Jodry (1999:329) argues that "the relative effects of bison hunting and mass processing on Folsom mobility, technology, and social organization were disproportionately great relative to the influences of other subsistence options." Such statements reveal a sentiment, perhaps one that is warranted, that subsistence concerns are primary among prehistoric hunter-gatherers and that other aspects of the cultural milieu are epiphenomenal—post hoc adaptations framed around the insurance of the production of a reliable food supply. Jodry (1999) argues that the Pleistocene extinctions coupled with the emergence of Younger Dryas climate created a boom in bison populations through nitrogen-enriched soils and forage, which set the stage for bison specialization by Folsom and Goshen hunter-gatherers. While a reliance on bison is well established in the core of the Folsom area, the western Great Plains and Rocky Mountains, no direct evidence of subsistence is available from peripheral areas including the eastern Plains and the Great Basin. In his study of Folsom projectile points east of the Mississippi River, Munson (1990) found that the geographic spread of Folsom points closely corresponds with the area defined by the historic Midwest Prairie Peninsula, suggesting to him that Folsom point users in the east also engaged in a bison-focal subsistence economy.

Although the ubiquitous presence of bison in Folsom faunal assemblages indicates regular use of this species, some authors have questioned whether Folsom subsistence was truly characterized by a specialization in large mammals (Amick 1994a; Kornfeld 1988; LeTourneau 1998). Similar arguments have been launched against the once pervasive notion that earlier Clovis subsistence strategies were geared almost exclusively toward the procurement of large mammals (e.g., Byers and Ugan 2005; Cannon and Meltzer 2004; Dixon 1999; Dillehay 2000; Meltzer 1988, 1993; Meltzer and Smith 1986). Furthermore, the faunal records of Paleoindian sites are not exclusively composed of large mammals, and there is evidence that Folsom foragers incorporated small game into their diets, including small mammals, reptiles, and possibly birds and fish (see Amick 1994a: table 5.2 and MacDonald 1998a:224 for tabulations of faunal assemblages from Folsom sites). For example, at Lindenmeier, Wilmsen and Roberts (1984:47) report, in addition to bison, evidence of the use of pronghorn, deer, hare, canids, and turtle by the Folsom occupants. Johnson (1987:124) argues from cut marks and consistent patterns of dismemberment that Folsom groups at Lubbock Lake utilized duck and muskrat. Recovered from Occupation 1 of the Downstream Locality of the Indian Creek site (believed to be Folsom) was evidence of the use of bighorn sheep, marmot, cottontail, and rodents (Davis and Greiser 1992:265).

Much of the debate concerning whether Paleoindians, in general, or Folsom hunter-gatherers specifically, were specialized hunters or generalized foragers depends on how the terms "specialized" and "generalized" are defined.

Following Waguespack and Surovell (2003), if the distinction is based on the degree to which prey types are not attacked upon encounter, then Clovis hunter-gatherers were, without a doubt, large-game specialists (see also Surovell and Waguespack in press). Being a specialist in this framework, however, does not preclude the use of small game, but instead suggests that small game will be taken only in certain circumstances (e.g., by individuals who are not able to hunt bison, at times when bison are scarce, or when small game encountered have low handling costs). In this framework, Folsom prey choice was closer to the "specialist" than the "generalist" end of the subsistence spectrum.

Folsom and Goshen bison hunting shows significant variation in the numbers of animals killed. Small bison kills involving fewer than five animals are known from sites such as Lubbock Lake (Bamforth 1985; Johnson 1987) and Waugh (Hofman and Carter 1991). Larger kills involving more than twenty animals are known from Mill Iron and Folsom, each with a minimum of just over thirty animals represented, and Lipscomb with fifty-five (Meltzer et al. 2002; Meltzer and Todd 2006; Todd et al. 1992; Todd et al. 1996), among others. Bison kills from this period show some variation in carcass density ranging from far less than one to approximately 1.3 individuals per $m^2$ (Hofman 1999b). Relatively dense bone beds suggest the use of natural topographic corrals, in particular arroyos, at sites such as Cooper and possibly Lipscomb (Bement 1997, 1999; Hofman 1999b; Todd et al. 1992). At more dispersed kill sites, such as Folsom and Upper Twin Mountain, the method used to dispatch the bison is less clear, but it is often argued to have also involved the use of natural landforms (Kornfeld et al. 1999:671; Meltzer 2006b:300–302; Meltzer et al. 2002:29). The use of a bison jump is inferred from Bone Bed 2 at Bonfire Shelter, containing both Plainview and Folsom artifacts (Dibble and Lorrain 1968), but this interpretation has recently been challenged (Byerly et al. 2005).

Bison hunting is known to have occurred year-round, supporting the notion that Folsom and Goshen peoples, like the historic hunter-gatherers who inhabited the same region, specialized in the procurement of bison. In contrast to the bone beds of later hunter-gatherers, however, Paleoindian bison bone beds do not show evidence of intensive butchery and marrow rendering (Todd 1991; Todd et al. 1990). At the Folsom site, for example, Meltzer et al. (2002:19) note that thorough disarticulation suggests the bison may have been extensively butchered, but: "There is virtually no evidence of bone impact fractures. Nor do any elements show signs of on-site processing for bone grease." Even less butchery is evident from other sites. The three kills at Cooper and the kill at Lipscomb, for example, contain highly articulated bison remains with relatively little evidence of butchery (Bement 1999; Todd et al. 1992). This "gourmet" butchery strategy, even for cold-weather bison kills, led Todd (1991:218) to suggest that Paleoindian hunter-gatherers appeared to be "fat indifferent," a surprising pattern given the progressively lean state of bison through the course of the cold season. He further argues that reduced late Pleistocene/early Holocene seasonality may have reduced the period of fat stress for bison and their Paleoindian predators, and that consumption of rumen contents could have provided an alternative source of carbohydrates in lieu of bone grease. Jodry (1999:329) makes a similar argument with specific reference to Folsom: "I suggest that greasing rose in economic importance after 10,000 BP due to deteriorating environmental conditions that affected bison forage and was coincident with a decrease in bison size (and average condition). It is proposed that subcutaneous and marrow fat reserves in bison during the early Younger Dryas period were substantially greater than in later Paleoindian times. . . . A higher incidence of subcutaneous and organ fat in Folsom age bison may be a significant factor affecting butchering and processing strategies." Although limited butchery at some Folsom sites supports the idea of fatty bison and organ feasts, the case for fat indifference may be somewhat overstated.

In his analysis of the Folsom component in Area 2 of the Agate Basin site, Hill (1994) found evidence that the butchery of both bison and pronghorn were geared toward fat retrieval. Based on cut marks and impact fractures, he argues, "The data suggest that Folsom hunters at Agate Basin sought only those elements and/or associated muscle or organ masses from each taxon with extremely high fat contents" (Hill 1994:126). The evidence from Agate Basin, believed to have been occupied in the late winter or early spring (the time of greatest nutritional stress for bison), conforms to the expected pattern of fat-conscious butchering for cold-weather occupations. A similar result was obtained from Byers's (2001) reanalysis of the bison remains from the Agate Basin (complex) level at Hell Gap, Locality II. In this assemblage, green bone breaks are common (found on 66 percent of long bones), and the frequency of breaks by skeletal element correlates with the marrow utility of elements. Like Hill (1994), Byers (2001:99) argues that "subsistence behaviors inferred from the Locality II Agate Basin bison remains suggest a feeding strategy designed to maximize the amount of fat included in the diet, most likely in response to a diet dependent on lean meat." Although seasonality data from this component suggest year-round hunting of bison, the most intensive use of the locality was during the fall, winter, and spring (Byers 2001:26–27). Therefore, Paleoindian sites do often show less intensive butchery than do Late Prehistoric plains bison kills, but evidence for marrow rendering is by no means absent.

*Mobility*

Renowned for the scale of residential mobility in which they engaged, Folsom hunter-gatherers "may have been the most mobile pedestrian hunter-gatherers to inhabit the globe" Amick (1996:411). This sentiment is derived from the oft-made observation that lithic raw materials present in Paleoindian assemblages, and particularly those of Folsom, have been transported over huge distances, commonly exceeding 400 km (e.g., Amick 1994a,b,c, 1996; Hester and Grady 1977; Hofman 1999a; Hofman 1991). This pattern is especially well documented for Southern Plains Folsom sites, but evidence for the long-distance transport of lithic raw materials is present in sites in the Northern Plains and Rocky Mountains as well (e.g., Frison 1982a, 1984; Jodry 1999; MacDonald 1999; Root 2000; Waguespack et al. 2002). As discussed above, Folsom mobility is then seen as an adaptation to the hunting of bison, a highly mobile, gregarious species.

Many of the current views of early Paleoindian mobility stem from Kelly and Todd's (1988) "Coming into the Country" model. They envision late Pleistocene foragers engaged in a "system in which residential groups moved from kill to kill" (Kelly and Todd 1988:236). Again, they point to limited data supporting heavy butchery and greasing as evidence of frequent residential mobility in pursuit of large game. In contrast to Late Prehistoric inhabitants of the Plains who heavily processed bison kills, in part for food storage in anticipation of future shortages, they argue that Paleoindians instead would "begin an almost immediate search for further resources" following a kill event (Kelly and Todd 1988:238). A reliable supply of bison (i.e., regular encounters with bison) would eliminate the need for intensive processing and result in serial residential moves over long distances.

A similar model for Folsom in the Southern Plains and Basin and Range has been proposed by Amick (1994a, 1996, 2000). Amick sees Folsom hunter-gatherers as mimicking the hypothesized east-west migrations of bison from the Southern Plains in the warm season to the sheltered basins of New Mexico in the cold season. Based on the regular occurrence of central Texas Edwards Plateau chert in Folsom sites in the New Mexico Basin and Range, Amick believes Folsom peoples regularly "geared up on Edwards chert sources, then exploited the grasslands of the Southern Plains during summer and early fall and then moved on to winter in the intermontane Rio Grande Valley" (Amick 1996:417). He suggests that the total annual residential mobility of some Folsom groups in this region could have exceeded a distance of 1,400 km (Amick 1996:418). Using ratios of projectile points to preforms and ratios of point bases to tips compiled by region, Amick (1994a, 2000) further argues that Folsom hunters logistically exploited the Southern Plains and small basins and uplands of the Basin and Range, while frequent residential moves and limited logistical forays were used in the larger Tularosa, Albuquerque, and northern Jornada del Muerto basins.

Hofman (1999a) suggests a similar pattern of movement for Southern Plains Folsom groups but instead argues for a redundant pattern of land use, independent of region. He further suggests that Amick's data may be problematic because of his heavy reliance on data from private collections, in which Folsom preforms are often underrepresented because they are not as readily identifiable as finished points (Hofman 1999a:401). Based on regional analyses of Folsom points from the Southern Plains, Hofman (1994, 1999a) notes consistent patterns of movement in lithic raw materials. In particular, Edwards chert, the raw material of choice for Southern Plains Folsom peoples, is most often transported to the north and northwest of the source area in central Texas, up to a distance of

600 km. Alibates silicified dolomite from the panhandle of Texas tends to have been moved north, northwest, and east, but notably it very rarely occurs in the form of Folsom points in southern Texas (Hofman 1994).

Hofman (1999a:403–407) proposes that Southern Plains Folsom bands acquired Edwards chert during the winter. He suggests that in the cold season, residential mobility was reduced while groups wintered in the dissected plains, off the high Llano Estacado, relying on "sheltered settings, diverse resources, and dispersed bison herds . . ." (Hofman 1999a:405). Winter was a time for hunkering down and replenishing and repairing toolkits. He does not, however, argue that winter residential mobility would have always been restricted. Depending on the responses of bison to climate, Folsom groups may have utilized all areas of the Southern Plains during all seasons. During severe winters, however, Folsom groups and bison would have retreated to more sheltered settings (Hofman 1999a:405–406; see also Greiser 1985). He finds support for this model in seasonality estimates for the Folsom, Lipscomb, Lake Theo, and Cooper sites, all late summer to early fall kills falling on the margins of the southern High Plains. (Hofman 1999a:404). Arguments for reduced cold season residential mobility have also been made based on excavations of large high-density Folsom campsites in the Rocky Mountains (Stiger 2006; Surovell 2003a; Surovell, Waguespack, Kornfeld et al. 2003; Waguespack et al. 2002).

While extreme long-distance transport (greater than 400 km) of lithic raw materials is common in the Southern Plains, it is less common on the Northern Plains (MacDonald 1999). Contrasting the Bobtail Wolf site in North Dakota, where less than one percent of artifacts are made on nonlocal sources, with Shifting Sands in Texas, where 99 percent of the stone is from 150 km away, MacDonald (1999:142) argues, "These contrasting patterns of regional exotic stone utilization suggest occasional large-scale Folsom group movements (or patterned individual movements) in the Southern Plains and infrequent individual large-scale movements in the Northern Plains."

MacDonald adds another variable to the equation here: the transport of stone by individuals versus that of groups. Using Beaton's (1991) dichotomy between "estate settlers" and "transient explorers," MacDonald (1999) allies Northern Plains Folsom groups with estate settlers wherein most movement occurred within a relatively small region and large-scale movement was done by individuals in seeking mates or maintaining social ties. In contrast, in the Southern Plains, he argues, the regular occurrence of high percentages of nonlocal raw materials indicate long-distance movement by groups. As is discussed in detail in the following chapter, noting a similar pattern of abundant local raw material in the Hanson assemblage from the Bighorn Basin of Wyoming, Ingbar (1994) suggests that such a pattern can be explained by similar mobility regimes operating in different raw material environments. Using Ingbar's (1994) model, more frequent encounter with lithic raw material sources on the Northern Plains would result in frequent retooling and the mean distance of transport of any given raw material would be significantly lower.

Summarizing, there is a general consensus that late Pleistocene foragers in the Rocky Mountains and western Plains regularly undertook long-distance movements on the scale of hundreds of kilometers in pursuit of nomadic bison herds. Points of disagreement center on questions regarding the presence of seasonal rounds, individual versus group mobility, and regional differences in mobility regimes. Whether Folsom and Goshen point users engaged in frequent residential mobility year-round, or seasonally settled into winter camps, remains to be determined. Can Folsom mobility patterns from the Southern Plains be superimposed onto other regions? Have nonlocal lithic raw materials entered sites by the hands of one or a few individuals, or by the long-distance movement of cohesive bands? It is not my intent to address all of these issues in this study, since I am more concerned with the general principles governing technological variability than with "ethnographic reconstruction" of the past, although this is to some degree unavoidable.

## *Chipped Stone Technology*

Studies of Folsom technology have generally taken two forms, either site-specific analyses, often couched in a regional perspective (e.g., Bement 1997; Boldurian 1990, 1991; Boldurian et al. 1987; Boldurian and Hubinsky 1994; Buchanan 2002; Frison and Bradley 1980; Jodry 1999; MacDonald 1998b, 1999; Nami 1999; Peterson 2001; Root et al. 1999; Sellet 1999; Tunnell 1977), or large-scale regional surveys of projectile points, and to a lesser extent preforms and channel flakes (Amick 1994a,b, 1995, 1996, 2000; Hofman 1999a; Naze 1986). The emphasis on projectile points in

Folsom technological studies, particularly at a regional scale, is warranted because Folsom points are temporally diagnostic. Folsom points have readily identifiable production stages and by-products, both in successful manufacture (channel flakes) and failure (failed preforms). However, of the 48,501 artifacts analyzed in this study, excluding surface finds, only 53 are projectile points or point preforms. As Bamforth (2002b) has forcefully argued, projectile points (0.1 percent of the study assemblage used here) have only limited utility in studies of Paleoindian lifeways.

In contrast to projectile point technology, Paleoindian core technology has received relatively little attention, with the possible exception of Clovis blade production (Collins 1999). Frison and Bradley's (1980:18) study of the Hanson assemblage defined three primary Folsom core types: discoidal, bifacial thinning, and opportunistic flake production. Discoidal cores are morphologically similar to Middle Paleolithic centripetal Levallois cores, in which a dorsal convexity is created for the production of large flake blanks (see Judge 1973:166). Discoidal cores have been recovered from at least two other Folsom sites, Barger Gulch, Locality B (Surovell, Waguespack, Kornfeld et al. 2001; Surovell, Waguespack, Richings-Germain et al. 2001) and Agate Basin (Bradley 1982:183). More common in assemblages, however, are informal cores (Bamforth 2002b), the result of what Frison and Bradley (1980:18) call "opportunistic flake production." For example, at Barger Gulch, Locality B, of the thirty-two cores recovered from excavation, twenty-four show unpatterned flake removals. This is repeated at Hanson (Frison and Bradley 1980) and Bobtail Wolf (Root 2000). Bifacial cores, however, are argued by Hofman (1992) to be the centerpiece of Folsom mobile toolkits, not only producing large tool blanks for flake production, but also serving as tools and preforms for projectile point production (an adaptation of Kelly 1988; Kelly and Todd 1988). A number of surface finds of oversize bifacial cores are known from the Southern Plains (e.g., Stanford and Broilo 1981; Wyckoff 1996), but they are not known from excavated Folsom assemblages, although flakes and flake tools from such cores have been recovered (e.g., Hofman et al. 1990; Jodry 1999:154). In contrast, LeTourneau (2001), found only limited evidence of unifacial tool production on bifacial blanks at Lindenmeier (28 percent of tools) and Blackwater Draw (23 percent of tools). Bamforth (2002b) likewise argues that bifaces played only a minimal role as cores in Paleoindian technology, and data from this study confirm these findings in that only 24 of the 333 (7.5 percent) retouched flake tools examined could be confidently determined to have been produced on flake blanks derived from bifacial reduction.

Of Kelly's (1988) three sides of a biface—core, long use-life tool, and a by-product of standardization—the first two have been proposed as explanations for the use of biface technologies by Folsom peoples. At least three formal types of bifaces are regularly recognized in Folsom assemblages, cores, projectile points (and preforms), and ultrathins. As discussed above, many authors have proposed that large bifacial cores were manufactured as a means of using raw material efficiently under conditions of low raw material availability (Bolduarian 1991; Hofman 1992; Jodry 1999; Root et al. 1999; Stanford and Broilo 1981; Wyckoff 1996). Ultrathin bifaces typically exceed a 10:1 width-to-thickness ratio and often exhibit a biconcave cross section (Jodry 1998, 1999; Root 2000; Root et al. 1999). These artifacts are so thin relative to their width that they can approach and even exceed the width-to-thickness ratios seen in ordinary bifacial thinning flakes. The notion of Folsom bifaces as long use-life tools has been applied both to Folsom projectile point production and ultrathins, again as a means of conserving raw material. Jodry (1998, 1999) suggests ultrathin bifaces may have served as specialized jerky knives for removing thin sheets of meat. Similarly Root et al. (1999:164) see ultrathins as "heavily curated" tools that are "designed for long use." Evidence from Lake Ilo sites in North Dakota suggests that Folsom ultrathins were also recycled by intentional radial fracture (Root et al. 1999), and a small fragment of an ultrathin from Barger Gulch, Locality B (avg. thickness = 4.5 mm) refits to a failed Folsom preform on a bend break facet, demonstrating that these tools sometimes were recycled into other tool forms (Surovell, Waguespack, and Kornfeld 2003). The production of Folsom projectile points is also commonly argued to have been a prolonged event, in which the biface is utilized at multiple stages to take advantage of its many renewable edges (Bolduarian and Hubinsky 1994; Ingbar and Hofman 1999; Judge 1973:168–169). Even the fluting of Folsom points has been argued to have been a means of controlling impact fractures, thereby maximizing the number of use events squeezed out of a single projectile point (Ahler and Geib 2000).

Although unifacially retouched flakes are the most common tool forms in Folsom and Goshen assemblages, there is no standardized typological scheme for classification of these artifacts. The number of typological frameworks employed approaches the number of studies of Folsom and Goshen tool assemblages. Bradley, for example, applied Bordes's Middle Paleolithic typology to the Folsom tool assemblages from Hanson (Frison and Bradley 1980) and Agate Basin (Bradley 1982), although he did not apply it to the Goshen assemblage from Mill Iron (Bradley and Frison 1996). Irwin and Wormington (1970) developed a typology for Paleoindian tool types based on repeated tool forms they observed in numerous Paleoindian components across the Great Plains. Perhaps the most rigorous attempt to develop a classification of Folsom tool forms was that of Wilmsen (1970: 51–62), wherein a simple clustering algorithm using the co-occurrence of tool form attributes was employed to define types. Despite these attempts to classify the variability seen in Paleoindian tool forms, all of these typologies seem to have been largely ignored since their publications.

Of the wide array of tool forms present in Paleoindian tool assemblages, only two, endscrapers and gravers, are universally recognized. Endscrapers exhibit steep invasive retouch on the distal edge, although retouch commonly is found along the entire tool edge. In outline, they typically flare outward distally to a wide distal edge. Folsom endscrapers are commonly spurred with a graverlike projection protruding from one distal edge. Unlike most Folsom flake tools, gravers are commonly produced on bifacial thinning flakes (LeTourneau 2001), likely because it is easier to produce a delicate graver tip on a thin flake edge than a thick one. Gravers are typically produced by isolating a pointed tip on a flake edge by the retouch of slight notches on one or two sides of the tip. Both single and multiple gravers are common. Some authors also recognize radial break tools, in which flakes (or bifaces) are intentionally fractured with a blow to the flake surface to create burin-like edges (Frison and Bradley 1980; Root et al. 1999). Many other flake tool forms are known but not consistently defined or recognized, including notches, burins, side-scrapers, drills, and retouched channel flakes.

The integration of these various stone tool forms into a dynamic picture of the Folsom/Goshen "technological system" is problematic. Derived from Spiess's (1984) use of the phrase, Kelly and Todd (1988:237–239) characterized early Paleoindians as "high technology foragers," meaning that many technological strategies were geared toward portability and conservation of lithic raw materials, such as the use of bifaces, high-quality lithic raw materials, and the design of tool forms intended to maximize use-life. However, it is likely that technological strategies varied by region, environment, and time (Amick 1994a,b, 1995, 1996, 2000; Bamforth 2002b; LeTourneau 1998; MacDonald 1999). Nonetheless, the basic tool, core, and biface forms discussed above are common to assemblages throughout the western Plains and Rocky Mountains. As with mobility and subsistence, however, significantly less information is available in the peripheral areas of the Folsom range, such as the Great Basin and eastern Plains, with the obvious exception of projectile point morphology.

Because Folsom mobility is commonly believed to be an adaptation to a subsistence strategy focused on the procurement of bison, it follows that Folsom technological strategies are largely seen as an adaptation to high mobility. The most recurrent theme throughout discussions of Folsom technology is raw material conservation. Frequent movement over large distances in pursuit of bison makes for an unpredictable raw material supply since the distribution of bison and tool stone may not be congruent. Therefore, technological strategies should be geared toward efficiency in raw material use. Raw material conservation is identified through extensive resharpening of tools (Ahler and Geib 2000; Hofman 1991), the use of bifacial cores (Hofman 1992; Stanford and Broilo 1981; Wyckoff 1996), recycling of broken tools (Ingbar and Hofman 1999; Wilmsen and Roberts 1984), and staging of projectile point manufacture (Boldurian and Hubinsky 1994; Ingbar and Hofman 1999). Another possible strategy for raw material conservation involves alternative methods of projectile point production (Amick 1994b; Hofman 1994; Stanford 1999). Because the fluting of Folsom points is a risky process often resulting in breakage of the preform, it is commonly argued that in contexts of raw material stress, the choice is made not to flute, thereby avoiding the possibility of wasting precious tool stone (Amick 1994b, 1995; Hofman 1991, 1992; see also Sellet 2004).

## The Study Sample

In this section, I provide a brief overview of the sites included for analysis in this study: Locality B of the Barger Gulch site, Upper Twin Mountain, Krmpotich, the Area 2 Folsom component of the Agate Basin site, and the Folsom component of Carter/Kerr-McGee. Lithic analysis included all artifacts recovered from excavations and in some cases diagnostic surface finds.

### Barger Gulch, Locality B, Colorado (5GA195)

The Barger Gulch site in Middle Park, Colorado, consists of at least ten Paleoindian localities spread over an area of approximately 3 km$^2$, situated adjacent to a perennial spring-fed tributary of the Colorado River, approximately 9 km southeast of the town of Kremmling (Waguespack et al. 2006). The erosional slopes flanking Barger Gulch, the stream from which the site derives its name, expose plentiful seams of Miocene Troublesome Formation chert in the site area (Izett 1968; Kornfeld and Frison 2000; Kornfeld et al. 2001; Naze 1986; Surovell, Waguespack, Kornfeld et al. 2001; Surovell, Waguespack, Kornfeld et al. 2003; Surovell, Waguespack, Richings-Germain et al. 2001; Waguespack et al. 2002; White 1999). Diagnostic artifacts from Barger Gulch span most of the Paleoindian period; finds include Goshen, Folsom, Hell Gap, and Cody projectile points (Kornfeld 1998; Kornfeld and Frison 2000). Folsom and Goshen are particularly well represented, being present at three localities each. Although the density of Younger Dryas archaeology in the Barger Gulch area is extremely high, this pattern is repeated throughout Middle Park where more than forty Folsom and Goshen localities are known (Kornfeld 1998; Kornfeld and Frison 2000; Kornfeld et al. 1999; Mayer et al. 2005; Naze 1986, 1994).

To date, excavations have focused only on Localities A and B, and of these only Locality B has proven to contain buried Paleoindian deposits (Kornfeld 1998; Surovell et al. 2000, 2005; Surovell, Waguespack, Kornfeld et al. 2001, 2003; Surovell, Waguespack, Richings-Germain et al. 2001; Waguespack et al. 2002, 2006; White 1999). Locality B (fig. 2.5) is a shallowly buried Folsom residential occupation dating to approximately 10,500 BP. The analysis includes all artifacts recovered through the 2002 field season. Through 2002, from 51 m$^2$, a total of 19,659 artifacts had been recovered, with an average density of 378 artifacts per m$^2$. The assemblage contains a wide range of artifacts, including projectile points, preforms, channel flakes, cores, spurred endscrapers, gravers, informal flake tools, and bifaces, including a fragment of an ultrathin biface (fig. 2.6a). Chipped stone artifacts are overwhelmingly dominated by local Troublesome chert ($\approx$99 percent). Other raw materials believed to be of local

FIGURE 2.5. Oblique aerial view of Barger Gulch, Locality B (marked by arrow), Middle Park, Colorado (looking north).

origin include a number of fragments of shattered quartz and a single flake of petrified wood. At least two nonlocal lithic raw materials are present. A yellow petrified wood, represented by a single tool (fig. 2.6h), is believed to be from the Black Forest source area east of the Front Range in the vicinity of Franktown and Elizabeth, Colorado. An orange dendritic chert, represented by more than two hundred items, compares favorably with Trout Creek jasper from the Arkansas Valley and South Park, Colorado. Although there is no direct evidence for season of occupation, Surovell, Waguespack, Richings-Germain et al. (2001) and Waguespack et al. (2002) suggest a possible prolonged cold season occupation where Folsom groups took advantage of bison herds wintering in the valley bottom.

### Upper Twin Mountain, Colorado (5GA1513)

Also located in Middle Park, Colorado, the Upper Twin Mountain site is a bison kill at 2,548 m above sea level on a high ridge just north of Little

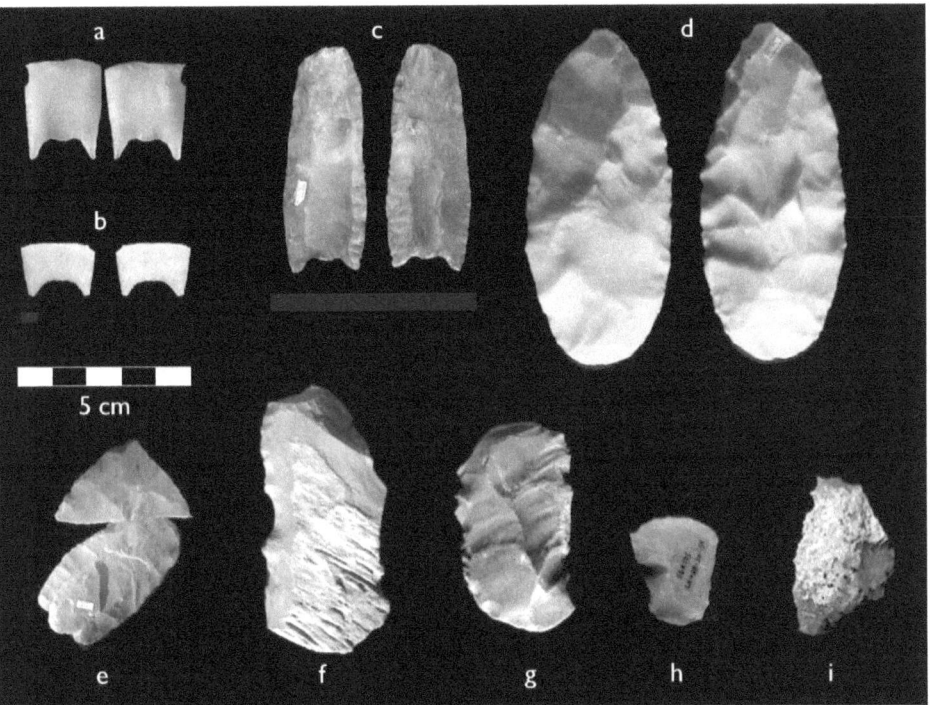

FIGURE 2.6. A sample of artifacts from Barger Gulch, Locality B. (a, b) projectile point bases, (c) complete fluted preform, (d) biface, (e) refit of a channel flake to preform fragments conjoining to an ultrathin biface fragment, (f) notched tool, (g) side-retouched flake tool, (h) endscraper, (i) graver.

Wolford Mountain (Kornfeld et al. 1999). The site preserves the remains of at least fifteen bison, shallowly buried in a remnant of a slump scar (fig. 2.7). Radiocarbon dates place the kill at approximately 10,300 BP. From dentition studies, the kill is estimated to have occurred in the late fall or early winter, and skeletal element profiles suggest the transport of high-utility portions away from the site.

In addition to bison bone, the 25.5 m² excavation produced eighty-seven artifacts, including two projectile points, five flake tools, and a core. The majority of the artifacts are small flakes less than 1 cm in maximum dimension. Two additional projectile points (fig. 2.8a, c) were recovered from the surface. All of the lithic raw materials from the site are local to the Middle Park area, but some are not local to the site (i.e., >20 km distant). Local raw materials in the assemblage include Troublesome

FIGURE 2.7. View of the Upper Twin Mountain Goshen bison kill (marked by white arrow), Middle Park, Colorado (looking northeast). Photograph courtesy of Marcel Kornfeld, University of Wyoming.

Formation chert, granodiorite, and a sedimentary stone, which Kornfeld et al. (1999:669) refer to as porcellanite. Nonlocal raw materials include gray Dakota Formation quartzite, available on the northwestern margin of Middle Park near Rabbit Ears Pass (Bamforth 2006) and a red to orange chert, possibly Table Mountain jasper, available primarily in the eastern half of Middle Park. The site is interpreted as a hunting locale for the "acquisition of surplus meat for permanent winter residence" in Middle Park (Kornfeld et al. 1999:672).

### Krmpotich, Wyoming (48SW9826)

The Krmpotich site, approximately 9 km southeast of the town of Eden, is a Folsom site in the Killpecker Dune Field of southwestern Wyoming, on the eastern edge of the Green River Basin (fig. 2.9). Discovered in 1969 by Jack Krmpotich, a local resident, the site contains buried Folsom artifacts vertically dispersed above and below a late Holocene disconformity within

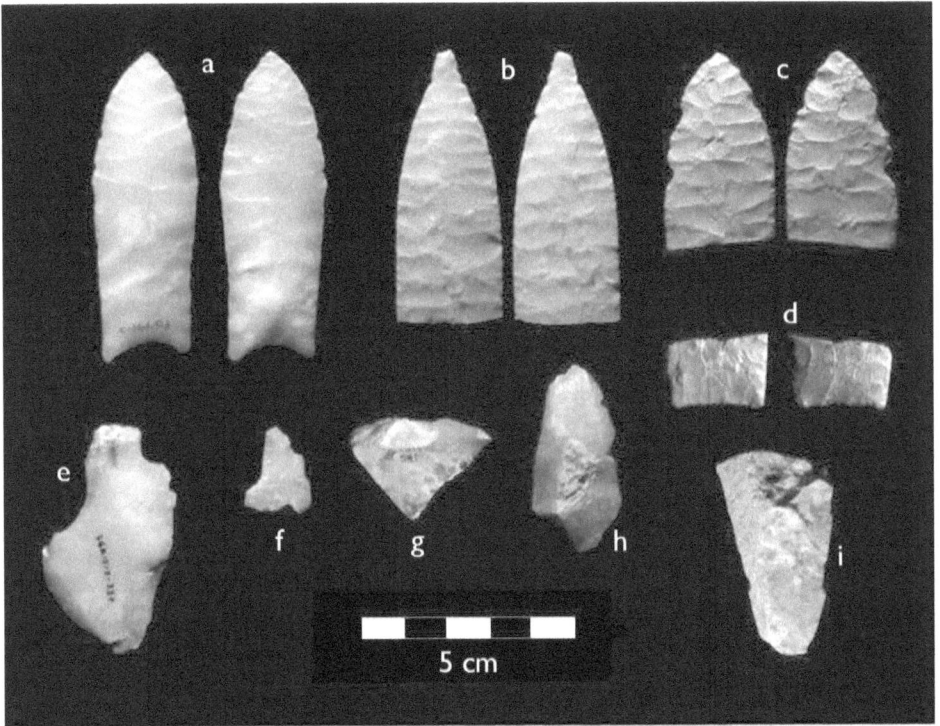

FIGURE 2.8. A sample of artifacts from Upper Twin Mountain. (a–d) projectile points, (e–i) flake tools.

a stabilized dune (Kornfeld et al. 1999; Mayer 2002, 2003; Peterson 2001). The excavated assemblage includes 8,866 chipped stone artifacts from an area of 74 m². My analysis of the lithic assemblage includes materials collected through the 2000 field season. In addition to thousands of small fragments of bone and tooth enamel, several identifiable specimens of bison bone are present from the site (Kornfeld et al. 1999), but their association with the Folsom occupation is unclear. Collagen from a well-preserved bison humerus, for example, produced a late Holocene radiocarbon age, not inconsistent with the age of the buried soil in which it was found (Mayer 2002). Chronological control is generally poor, with luminescence ages of 14,690 ± 870 and 1,500 ± 110 cal yr BP bracketing the occupation (Mayer 2002).

The site assemblage includes projectile points, preforms, channel flakes, bifaces, cores, and a range of tool forms including endscrapers and gravers

50  Chapter 2

FIGURE 2.9. View of the year 2000 excavations at the Krmpotich Folsom site, Sweetwater County, Wyoming (looking southeast). Photograph courtesy of Marcel Kornfeld, University of Wyoming.

(fig. 2.10). A variety of local raw materials are present in the assemblage, including several cherts derived from the Eocene Green River Formation, quartzites from the Farson gravels underlying the dune field, and a lavender silicified silt or mudstone. Nonlocal raw materials include various cherts, porcellanites, and obsidian (see Peterson 2001 for detailed raw material source information). Based on the diversity of tool forms present, the site is believed to represent a residential occupation or campsite (Kornfeld et al. 1999; Peterson 2001).

### *Agate Basin (48 NO201), Area 2, Folsom Component, Wyoming*

The Agate Basin site was an area of repeated use from the latest Pleistocene through the early Holocene. Among the many Paleoindian components recognized from the site's many localities are Clovis (Goshen?), Folsom, Agate Basin, and Hell Gap. The site, located in an intermittent tributary of the Cheyenne River in far eastern Wyoming (fig. 2.11), has been investigated off and on since the early 1940s (Agogino and Frankfurter 1959;

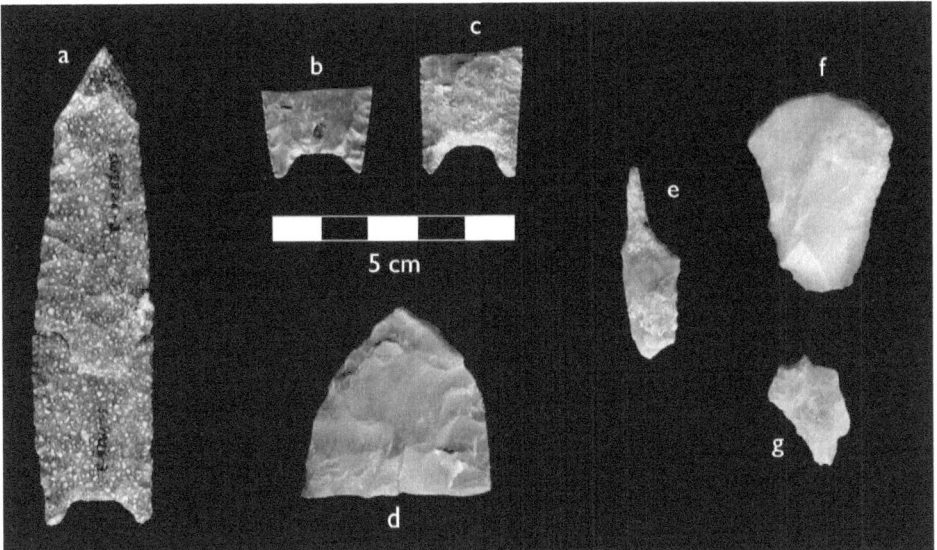

FIGURE 2.10. A sample of artifacts from the Krmpotich site. (a) projectile point preform, two refitted fragments, (b–c) projectile point bases, (d) biface fragment, (e) channel flake drill, (f) endscraper, (g) graver.

Bass 1970; Frison and Stanford 1982a; Roberts 1943, 1961; Sellet 2004). For this study, 17,907 artifacts were analyzed from the Folsom component of Area 2 (fig. 2.12). Area 3, approximately 80 m to the west, produced two distinct Folsom components, but Area 2 produced only one. Artifact refits and conjoins between the two areas indicate the contemporaneity of the occupations, at least with respect to one of the Area 3 components (Frison 1982b:38). A radiocarbon date of 10,780 ± 120 BP (SI-3773) provides an age estimate for the occupation (Frison 1982c:179).

The faunal assemblage from the Area 2 Folsom component contained the remains of pronghorn, peccaries, canids, rabbits, and at least eleven bison (Frison 1982b; Hill 1994, 2001). Based on bison dentition, the season of occupation is inferred to be late spring or early winter (Hill 1994, 2001). The bison and pronghorn assemblages are consistent with a residential occupation to which high-utility portions, or complete animals in the case of pronghorn, were transported, and as discussed above, butchery patterns are consistent with marrow extraction (Hill 1994, 2001). One or two structures are postulated to have been present in Area 2, but their functions are unclear (Frison 1982b:39–44; Frison and Stanford

FIGURE 2.11. Aerial view of the Agate Basin site, Niobrara County, Wyoming (looking southeast). White arrow shows the location of Area 2. Photograph courtesy of George Frison, University of Wyoming.

1982b:364). A poorly defined circular area, centered on a shallow hearth, contained a concentration of tools and flakes, and two ribs at its periphery are suggested to be tent stakes. Although Frison and Stanford (1982b:364) noted that the evidence for the existence of structures is "weak," using patterns of lithic refits Hill and Sellet (2000) found support for the existence of at least one structure in this component.

The lithic assemblage from Agate Basin contains the full range of Folsom tool and core forms. Projectile point technology is represented by finished points, preforms, and over a hundred channel flakes. Based on the analysis of lithic nodules involved in projectile point manufacture, Sellet (2004) argues that Folsom hunter-gatherers at Agate Basin "geared up" by manufacturing large numbers of projectile points that were transported away from the site in anticipation of future needs. In addition to projectile points, a large variety of flake tool forms were recovered, as were cores and bifaces. One quartzite biface fragment, with a width-to-thickness ratio of 9.3, could easily qualify as an ultrathin (fig. 2.12d).

Late Pleistocene Foragers 53

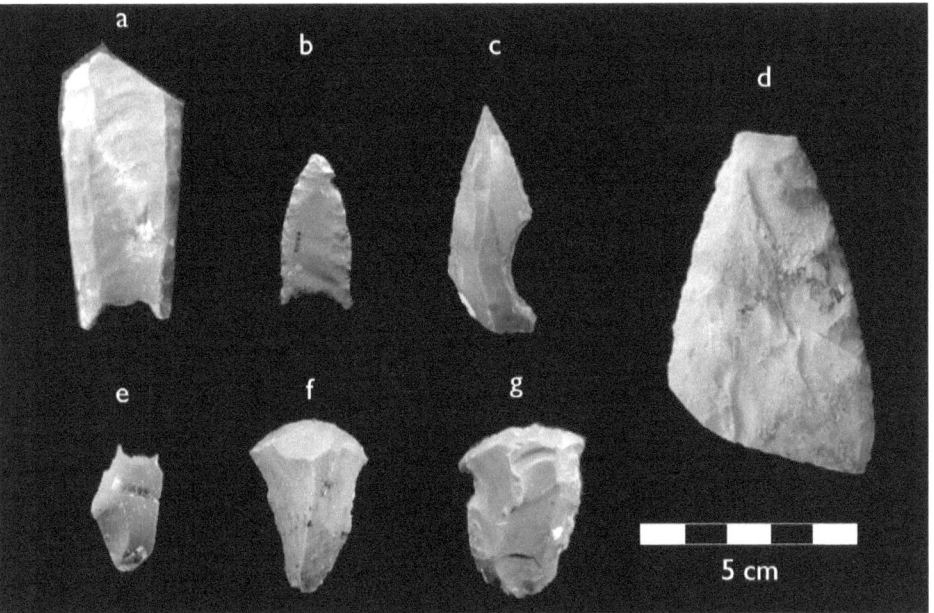

FIGURE 2.12. A sample of artifacts from the Agate Basin, Area 2, Folsom component. (a) fluted point, (b) pseudo-fluted point, (c) pointed and notched flake tool, (d) ultrathin biface, (e) graver, (f–g) endscrapers.

The assemblage is overwhelmingly dominated by small debitage recovered with 1/16-inch screens. An impressive variety of lithic raw materials are found in the assemblage (Craig 1983; Frison 1982a). Materials from the immediate site area include moss agate, petrified wood, and cobble cherts. A large variety of nonlocal raw materials are present as well, most notably Knife River flint, available approximately 500 km to the northeast. Hartville chert and Dakota Formation quartzite were moved to the site from the southwest, and White River Group chert from the south or southeast. An opaque red chert may have come from the Triassic Spearfish Formation in the Black Hills to the northeast.

## Carter/Kerr-McGee (48CA012), Folsom Component

The Carter/Kerr-McGee site is a multicomponent Paleoindian site in the Powder River basin of northeastern Wyoming. The site is adjacent to an intermittent stream within 100 m of an active spring (fig. 2.13).

FIGURE 2.13. View of the Carter/Kerr-McGee site, Natrona County, Wyoming (looking south). White arrow shows the location of the excavations. Photograph courtesy of George Frison, University of Wyoming.

Four components were recognized at the site; the oldest was initially believed to be Clovis (Frison 1984; Reiss et al. 1980) and has since been revised to Goshen (Frison et al. 1996). A Folsom level was overlain by a mixed Hell Gap–Agate Basin component. The uppermost cultural level was a dense Cody-aged bison bone bed containing Eden, Scottsbluff, and Alberta projectile points. Frison (1984) suggests that the site was a

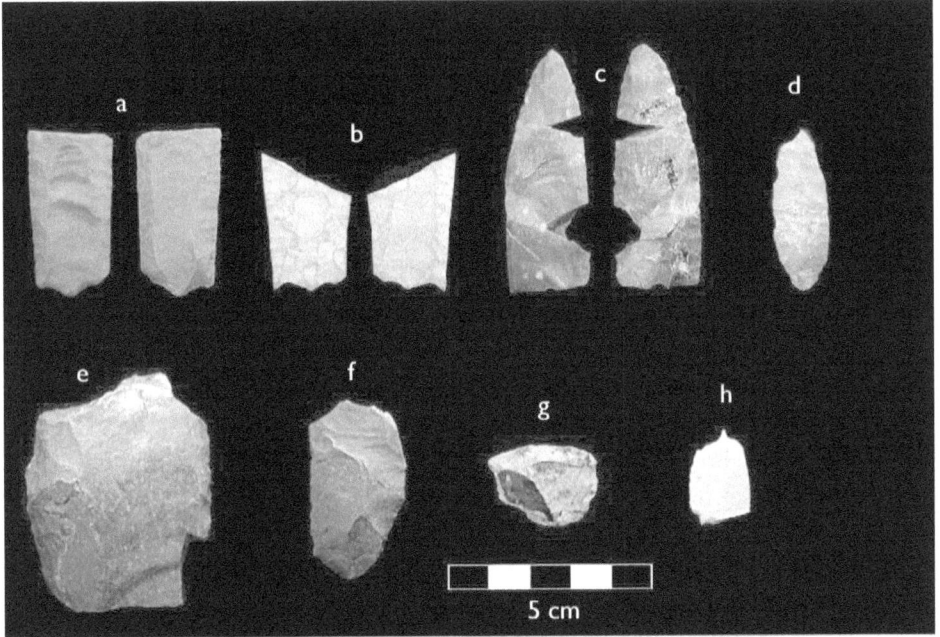

FIGURE 2.14. A sample of artifacts from the Carter/Kerr-McGee Folsom component. (a) projectile point (surface find), (b) projectile point, (c) preform, (d) channel flake, (e–f) flake tools, (g) endscraper, (h) graver.

natural bison trap that was repeatedly used throughout the Paleoindian period.

Although the faunal remains from the Folsom component are poorly reported, bison are present in the assemblage, and Frison (1984:303) suggests that at least one tibia was fractured to render marrow. A charcoal sample from a hearth produced a radiocarbon date of 10,400 ± 600 $^{14}$C yr BP (RL-917), but its imprecision makes it somewhat unreliable. The Folsom lithic assemblage totals 1,988 items, mostly tiny flakes recovered from 1/24-inch screens (Frison 1977 in Reiss et al. 1980). The assemblage includes projectile points, preforms, channel flakes, endscrapers, gravers, flake tools, and a single core (fig. 2.14). Less than 1 percent of the artifact assemblage is composed of local raw materials, which include petrified wood and porcellanite. Provenance of two nonlocal raw materials, Morrison quartzite and a red opaque chert (Phosphoria or Spearfish Formation), could be attributed to sources in the Black Hills to the east and/or the Bighorn Mountains to the west. A yellow to red opaque dendritic chert may have originated from sources in the Hartville uplift or the Black Hills (Rick Weathermon, personal communication, 2000). Also present in the assemblage are quartzites, most likely from sources in the Black Hills. The diverse tool assemblage and presence of a hearth suggests that the Folsom component of the Carter/Kerr-McGee site represents a residential occupation.

## Summary

The hunter-gatherers who lived in the Great Plains and Rocky Mountains during the latest Pleistocene were undoubtedly specialized bison hunters. It is argued that from this bison-hunting adaptation, a suite of economic strategies arose. Annual residential movement over large distances provided the mechanism for maintaining regular access to bison. Mapped onto high residential mobility was a toolkit designed for efficiency in raw material use to bridge long periods away from sources of high-quality tool stone.

The data set developed for this study is adequate for exploring technological variability, particularly with respect to the impact of mobility on human behavior as seen through archaeological assemblages. Put simply, extremes in mobility should result in extreme constraints on technology. The five sites incorporated into this study occur over a very large area and exist in drastically different circumstances. Upper Twin Mountain and Barger Gulch sit in the intermontane basin of Middle Park, surrounded on three sides by mountain passes approaching or exceeding 3,000 m. The Krmpotich site in southwestern Wyoming lies in the midst of the Killpecker Dune Field. The Agate Basin and Carter/Kerr-McGee sites are found in the shortgrass prairies of eastern Wyoming. The sample includes both camp and kill sites, and the nature of the site assemblages varies dramatically. Barger Gulch contains a high density of artifacts and is dominated by local raw materials, and in contrast Carter/Kerr-McGee exhibits a considerably lower artifact density and is dominated by nonlocal raw materials. I consider these facts to be assets to the study. People who used stone tools, regardless of time or environment (social, economic, natural, etc.) had by necessity to overcome the same basic problem of getting functional stone tools to the places they used them for the purposes for which they were intended to be used. In this sense, there is no reason why it is even necessary to hold time and place constant, except perhaps that it facilitates intersite technological comparisons by holding constant certain strategies of reduction. This study differs from most previous studies of Paleoindian technology in that the sample is not limited to projectile point technology nor to a single site, and in this respect it allows the exploration of variation within Paleoindian lifeways, which all too often are portrayed as being largely invariant, a sin of which I am somewhat guilty in this very chapter.

# 3

# Occupation Span and Residential Mobility

CONSIDER A SURFACE scatter of lithic artifacts, resting on bedrock, lacking diagnostics and patination. Does this assemblage represent the accumulation of a single or multiple occupations, and what was the mean length of the occupations represented? Is it even possible to answer these questions using only attributes of the lithic assemblage, without the aid of stratigraphic superposition and any indicators of age, or even spatial patterning? In short, I think the answer is yes. Of course, even when artifacts are recovered from buried contexts, the problem remains. Vertically constrained and stratigraphically identifiable cultural levels could also represent single or multiple occupations. My goal in this and the following chapter is to develop formal models of artifact accumulation in order to derive archaeological measures of occupation span and reoccupation of sites using only attributes of lithic assemblages. Because occupation span and the frequency of residential mobility are inversely related, measures of occupation by their very nature are also measures of mobility. These measures of mobility can then be used as independent variables for the investigation of technological variability.

Identifying the number and duration of occupations has been central to interpretations of occupation history and Folsom mobility at the Hanson site. Faced with an assemblage composed predominantly of local raw materials, Ingbar (1994:54) noted that an interpretation of mobility based solely on raw material representation would make the group who inhabited Hanson "one of the least mobile Folsom groups ever found." Because Hanson did not fit the typical pattern of an assemblage dominated by lithic raw materials transported long distances, Ingbar (1992, 1994) used a computer simulation and patterns of raw material differentiation among artifact classes to argue that Hanson instead was a quarry site that had been briefly occupied many times. Could the same evidence be used to argue that the Hanson site represents one or a few relatively long-term occupations and reduced mobility?

Hanson is a large Folsom site in the western Bighorn Basin of Wyoming. Two areas have been investigated, and Frison and Bradley (1980:10) estimate that 5 percent of the site has been excavated; it is not known how much of the site has been lost to erosion. A distance of approximately 45 m separates areas 1 and 2. Published reports inventory over 12,000 artifacts of which more than 500 were tools (Frison and Bradley 1980; Ingbar 1992). It is estimated, however, that well over 100,000 artifacts have been recovered from the excavations, most of which remain unanalyzed. While bone preservation was generally poor, the remains of at least three bison were recovered in addition to bones of mountain sheep, deer, marmots, and cottontail rabbits (Frison 1978; Frison and Bradley 1980). Frison (1978; Frison and Bradley 1980) believes that structures may have been present: "Hard-packed, sandy floors define what are believed to be vaguely circular living structures although no discernible postholes outline the floors" (Frison 1978:118). In a few locations, it was possible to distinguish two separate Folsom levels, but over most of the site they were indistinguishable.

Dominating the assemblage is Morrison chert, available within 1 km of the site. Locally available Morrison quartzite is also quite abundant. Other raw materials include Phosphoria (20 km) and Madison chert (40 km). Interestingly, projectile points and endscrapers are preferentially made on Phosphoria chert, leading Ingbar to conclude that they were transported to the site during residential moves preceding occupations. In contrast, Frison and Bradley (1980:113) suggest that Phosphoria chert was simply preferred for the production of these tool types. Amick (1994a:318) argues that, because Phosphoria chert is located only 20 km from the site, it was likely procured by individuals or task groups working out from the site.

In contrast to Ingbar's interpretation of the site as a quarry or workshop locality, Frison (1978:145) argues, "The present evidence suggests a campsite . . . [and] the cultural group may have been living there during the late fall or early winter months." Frison and Bradley (1980:130) also argue that the Hanson site was associated with a bison kill but admit that the current evidence for such is weak: "the Hanson Folsom site can be interpreted as a relatively large concentration of several families (possibly a band) who engaged in acquiring large amounts of stone flaking materials. The amount and nature of the faunal remains also suggest proximity to a reliable source of bison." While all authors acknowledge multiple occupations, Frison, Bradley, and Amick suggest very few are represented, possibly two or three. Ingbar (1992:186), on the other hand, sees Hanson as the result of "many occupation events."

In addition to Hanson, a number of other Folsom sites show similar patterning that contrasts with the "typical" Paleoindian pattern of low-density sites dominated by nonlocal raw materials transported over large distances. Sites like Barger Gulch, Locality B (Surovell, Waguespack, Kornfeld et al. 2003; Surovell, Waguespack, Richings-Germain et al. 2001; Waguespack et al. 2002, 2006), Hell Gap (Irwin-Williams et al. 1973; Sellet 1999), Adair-Steadman (Tunnell 1977), Mountaineer (Stiger 2006), the Lake Ilo sites (Root 2000; Shifrin 2000; William 2000), and possibly Lindenmeier (Wilmsen and Roberts 1984) are dominated by local raw materials and characterized by relatively high artifact densities. Whether sites like Hanson are the result of many short-term occupations or one or a few long-term occupations is critical to the reconstruction of Folsom settlement patterns. In particular, are large, high-density Folsom sites reoccupied quarry or workshop locales (Ingbar 1992, 1994; Stanford 1999:302), and/or do they represent one or a few relatively long-term residential occupations? The models developed in this and the next chapter point to ways that these questions might be answered.

## What Is Mobility?

Hunter-gatherers use mobility to position themselves within landscapes for the acquisition of natural and cultural resources (Binford 1980; MacDonald and Hewlett 1999). Though the term "mobility" in a vernacular sense relates to how people "move around," it encompasses a broad range of behaviors. Fundamentally, a move by a person or persons can be represented as a vector, described by a direction and a magnitude (distance). We can also attach a rate to that vector, essentially the distance moved per unit time. Any or all of these components of mobility could be studied with respect to individuals or groups. Alternatively, mobility could be used to refer to the dispersal or aggregation of individuals, or we could devise typologies of moves (e.g., Binford 1980), recognizing that different moves occur for different reasons.

In this study, I am concerned primarily with the frequency of residential mobility, the movement of individuals or groups of individuals from one residential base camp to another (Binford 1980). The frequency of residential mobility depends largely on subsistence and environmental variables (Binford 1980; Kelly 1983, 1995:111–160; Surovell 2000). Hunter-gatherers who rely primarily on hunted resources tend to move their base camps more frequently than those who depend increasingly on plants (Kelly 1995), and as environments become increasingly patchy, fewer residential moves by hunter-gatherers is an optimizing solution (Binford 1980; Kelly 1995; Surovell 2000). Population density can also condition mobility, since increased numbers of people on a landscape translate to fewer mobility options (Binford 2001; Kelly 1992). The degree to which hunter-gatherers are residentially mobile affects nearly all dimensions of human behavior, including demography, social and political organization, subsistence, technology, and ideology (reviewed by Kelly 1992, 1995).

## Formal Approaches to Monitoring Mobility Archaeologically

One very common approach to studying mobility (or lack thereof) is the use of formal models that attempt to describe the factors governing the creation of archaeological assemblages, what Varien and Mills (1997) have called "accumulations research." Two related strategies for modeling artifact accumulation are employed, computer simulation and formal mathematical models (e.g., Aldenderfer 1981; Ammerman and Feldman 1974; David 1972; de Barros 1982; Kintigh 1984; Kohler 1978; Kohler and Blinman 1987; Mills 1989; Schiffer 1975a,b; Schlanger 1990; Varien and Potter 1997). In this section, I limit discussion of accumulations research to the question of using archaeological assemblages to estimate site occupation span. The duration of time that a site is occupied is the reciprocal of the frequency of residential mobility. Simply put, long-term occupations translate to low frequencies of relocating home bases because time spent camping is time spent not moving.

In the 1970s two equations were independently developed to describe the primary factors affecting artifact discard. Both equations can be solved for occupation span, allowing estimates of site occupation duration based on artifact counts. In an ethnoarchaeological study of Fulani potters, David (1972:142) estimated the number of a particular type of pots discarded ($d_t$) as a function of time with the equation:

$$d_t = S_o + \frac{S}{2}\left(\frac{t}{L}\right) \qquad (3.1)$$

where S is the number of pots in use ($S_o$ is the number of pots in use at the start of the occupation), $t$ is time, and $L$ is the median use-life of those pots.[1] The equation was generalized by Deboer (1974:37) and later modified slightly by de Barros (1982:310). The original version states that the number of pots discarded as a function of time is the number in use at the start of an occupation (the first term on the right side of the equation) plus the number discarded during that occupation (the second term). The latter term is composed of $S/2$, the number of pots discarded within one use-life, and $t/L$, the number of use-lives elapsed during the occupation span.

Contrasting David's (1972) discard equation with that of Schiffer (1975a:840, 1987:53), clear contradictions are evident. Schiffer used essentially the same variables, but in a much simpler formulation:

$$d_t = \frac{S}{L}t \qquad (3.2)$$

where $S$ is the number of artifacts in systemic context, $L$ is the mean artifact use-life, and $t$ is time or occupation span. Schiffer's equation differs from David's in two important ways. It eliminates the first term in David's equation representing the number of artifacts in use at the start of the occupation. While David assumes that all of the artifacts in use at the start of an occupation will be discarded at that site, Schiffer makes no such assumption. Also, the second term in David's equation is identical to Schiffer's equation, but David divides the term by two. The differences may seem slight, but in fact the two equations can have drastically different solutions. This difference arises, in part, from the distinction between David's use of *median* use-life and Schiffer's use of *mean* use-life. Nonetheless, for artifacts where median and mean use-lives are equal (use-lives have a symmetrical distribution), both equations cannot be correct. Using an example from David (1972:tables 1 and 2), 33 bowls were in use at the time of his inventories, with a median use-life of 2.7 years. From Schiffer's model, 300 pots would suggest an occupation span of approximately 24.5 years, while David's would suggest duration of approximately 43.7 years. Without getting bogged down in a discussion of symmetry in distributions of artifact use-lives (but see Shott and Sillitoe 2004), Schiffer's equation is more appropriate because, on average, during a time span equivalent to one use-life, the total systemic inventory of a particular artifact type should be discarded, not one-half of that inventory as implied by David. Furthermore, mean use-life should have a consistent relationship to discard rate, while the relationship between median use-life and discard rate cannot be modeled without specifying a specific form of use-life distribution.

Using Schiffer's discard equation, Schlanger (1990) simulated the accumulation of archaeological assemblages as a function of occupation span assuming discard is a probabilistic process. Schlanger found that artifact type frequencies will be highly variable for short-term occupations, but as occupations are lengthened, artifact type ratios will stabilize. This property arises because the probability of discard of artifacts with relatively long use-lives is low for short-term occupations (Schiffer 1987:55). Also, Schlanger confirms what Schiffer (1987:54–55) calls the "Clarke effect," or the "statistical tendency for the variety of discarded artifacts to increase directly with a settlement's occupation span." Not surprisingly, long-term occupations should contain a greater diversity of artifact forms than short-term occupations. Also, the makeup of artifact assemblages (e.g., tool type frequencies) in short-term occupations may not reflect site function but instead the chance-driven probabilistic process of discard.

In a similar vein, Varien and Potter (1997) simulated the breakage and discard of cooking pots. They applied various assumptions about the nature of the systemic inventory of pots at the start of an occupation, the use of pots, the timing of artifact replacement, and the number of households. They found that the number of pots discarded is highly variable for short-term occupations, particularly those with few households. They demonstrate that as occupation span is lengthened and site population is increased, the results of their simulations converge with estimates of discard based on Schiffer's discard equation. As in Schlanger's work, when discard is modeled as a probabilistic process, short-term occupations are highly variable in assemblage content, but Varien and Potter note that while time is a key variable, *person time*, the cumulative time spent for all site occupants, is ultimately controlling the numbers of artifacts discarded.

They also note that use-lives and systemic inventories may not be independent since the use of greater numbers of a particular artifact type allows wear and tear to be spread between tools, effectively reducing the probability of breakage or discard for any given item.

While most accumulations research has emphasized the estimation of occupation span for single occupations, relatively few studies have attempted to derive methods for approximating the mean occupation span of reoccupied sites. Kohler and Blinman (1987) develop methods for doing one better— solving for multiple occupation spans in mixed assemblages using least-squares regression analysis. This technique apportions assemblages into phases, allowing estimates of occupation span for each phase, assuming population size for each phase is known. This method is of limited utility since it requires that population sizes are known and that a calibration set, or the expected frequencies of artifact types within each phase, is in hand. Although it may have broad utility to ceramic assemblages where regional sequences and seriations of ceramic types are well established, it cannot be used when such information is not available, as would be the case for many preceramic sites. Also, when multiple occupations are represented within phases, the coarse resolution of the technique would not allow the discrimination of artifact sets or occupation spans from different occupations.

In a study of Mogollon pithouse sites, Lightfoot and Jewett (1984, 1986) developed methods for detecting reoccupation, and Gallivan (2002) applied similar methods to a set of late precontact- and contact-era sites in the James River Valley in southeastern Virginia. These studies make a distinction between *residential stability* and *use duration of an archaeological place*. Residential stability is defined as: "the length of time spent in one site on an annual basis," and use duration is defined as "the aggregate number of years or time that an archaeological place is either continuously or repeatedly occupied" (Lightfoot and Jewett 1984:49). Using these two variables, a framework is developed for differentiating between sites composed of single occupations, whether seasonal or year-round, from those representing multiple occupations (fig. 3.1). There is some confusion, however, with respect to sites characterized by high residential stability. Lightfoot and Jewett (1984:50) suggest that sites characterized by high residential stability and use duration represent "year round villages continuously occupied for an extended duration," while Gallivan

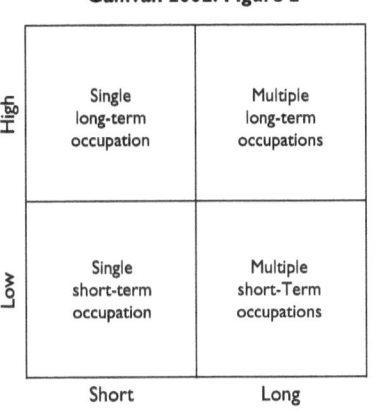

FIGURE 3.1. The Gallivan (left) and Lightfoot and Jewett (right) models for distinguishing between single and multiple occupations and short- and long-term occupations.

(2002:fig. 2) argues they represent "multiple long-term occupations." The contradiction is in the number of occupations represented. Whereas Lightfoot and Jewett believe that such sites are characterized by a single occupation, Gallivan sees multiple occupations.

I suggest that the confusion derives not from the upper right cell, but instead from the upper left cell characterized by short use duration and high residential stability. In reality, the conditions described by this cell cannot exist *if the x- and y-axes are equally scaled*. Using Lightfoot and Jewett's (1984:49) definitions, it is impossible for "the length of time spent in one site on an annual basis" (the residential stability) to exceed "the aggregate number of years or time that an archaeological place is either continuously or repeatedly occupied" (the use duration). Residential stability can only be equal to or less than use duration.

This problem aside, the methods used by Gallivan (2002) and Lightfoot and Jewett (1984, 1986) deserve further attention. To apply these models archaeologically, it is necessary to derive indices, or proxy measures, of residential stability and use duration. Lightfoot and Jewett use the percentage of pithouses containing interior hearths as an indicator of residential stability because residences occupied during the winter or year-round on the high Mogollon Rim should contain interior hearths.

In contrast, pithouses occupied during only the summer months, they argue, should lack indoor hearths. Therefore, they argue that low frequencies of interior hearths should indicate relatively low residential stability. As an indicator of residential stability, Gallivan (2002:542) uses feature diversity, because increased diversity in feature types implies that "a greater range of activities resulted from a residential presence that incorporated a longer portion of the settlement round." Next, a bivariate correlation matrix of variables is created to determine which aspects of assemblages, architecture, and/or features statistically correlate with the initial measure. For example, Lightfoot and Jewett (1984:53) find that the diversity of features, the mean number of postholes per house, the mean number of total interior features (including postholes), and artifact diversity correlate significantly with the percentage of interior hearths. The residential stability index uses the mean rankings of a set of sites for all independent, correlated variables.

A similar procedure is used to create a use duration index. Lightfoot and Jewett (1984:56–57) use artifact density, ratio of burials to pithouses, size and depth of middens, and the rate of superimposition (the number of features per site that show signs of remodeling or superimposition) as measures of use duration. Similarly, Gallivan (2002:543) uses a composite of feature density, postmold density, burial density, artifact density, and burial to house ratio.

These studies are important attempts to address a difficult and persistent archaeological issue. However, they are difficult to apply because of theoretical and methodological problems. In particular, one must accept that the initial or seed variable indicating residential stability, percentage of houses with interior hearths for Lightfoot and Jewett (1984, 1986) and the diversity of features for Gallivan (2002), does in fact reflect residential stability. If it does not, then the entire analysis fails, because the remaining variables are argued to monitor residential stability only if they correlate with the seed variable. It is reasonable to question whether Lightfoot and Jewett's use of percentage of houses with interior hearths is truly an indicator of residential stability; after all, they admit that it should in part be an indicator of seasonality. If so, then a site with 100 percent of the pithouses containing interior hearths could indicate year-round occupation, as the index would imply, or only a winter occupation, the opposite of what the index would imply. In this and the following chapter,

I develop formal models that are conceptually similar to those of Lightfoot and Jewett (1984, 1986) and Gallivan (2002) for exploring the problem of occupation span and reoccupation.

## Dissecting the Occupation Span Concept

It would be a difficult task to derive the frequency distribution of occupation spans for a group of prehistoric hunter-gatherers from a single site. Sites provide us with location-specific glimpses into settlement organization, and as such, they represent only a limited time slice of a mobility regime. Certain aspects of assemblages may provide information regarding the nature of previous residential moves, or of planning for future moves (e.g., Ingbar 1994; Larson 1994), but a regional approach combining a number of sites is superior to any site-specific approach, because it allows one to put individual sites in context and estimate ranges and distributions of occupation spans that are critical to describing prehistoric mobility strategies.

There are a number of mitigating factors that must be controlled before rigorous estimates of site occupation span are possible. There are demographic issues such as the number of individuals occupying a site and the stability of group composition. Hunter-gatherers can have notoriously flexible social organizations, with band membership changing almost daily (Binford 2001; Lee 1979; Steward 1938; Woodburn 1968; Yellen 1977). From the perspective of a single place, the composition and number of individuals present through time may be constant, or in a state of constant flux. Second, as discussed above, archaeological components are often treated as single occupations for interpretative simplicity. The possibility of multiple spatially overlapping occupations that are stratigraphically inseparable raises the possibility that measures of occupation duration may be monitoring the cumulative products of multiple occupations. These issues are well known, but they beg for a clear definition of occupation span since it may vary dramatically for each site occupant and each site occupation.

I begin with three separate measures of occupation duration. *Occupation span* is defined as the time elapsed from the arrival of the first occupant at a site to the departure of the last occupant for any continuous occupation. This is contrasted with *occupation intensity*, which is the sum of all time spent at a site for all inhabitants. Occupation intensity is measured as the sum of all time spent at a site by all occupants in a unit of *person time* (e.g., person days). Finally, mean per capita occupation span is the average length of stay per site occupant. Comparing these measures for hypothetical site occupation histories shows how they permit one to sort out the problem of changing group membership. For example, a site occupied by 30 people for 30 days has an occupation span of 30 days and an occupation intensity of 900 person days. If group membership remained constant (no one left or entered the group), the mean per capita occupation span would be 30 days. On the other extreme, if every day the group of 30 was replaced by a new group of 30, the mean per capita occupation span would be one day.

To illustrate how different aspects of an assemblage will reflect these measures of occupation duration, I will begin with a simple model of artifact discard. Assume a group of people stays at a site for a time ($t$), and group membership remains constant (i.e., no individuals leave or enter the group). The number of artifacts discarded ($d$) can be modeled as:

$$d = prt \qquad (3.3)$$

where $p$ is the number of site occupants, $r$ is the per capita discard rate (the number of artifacts discarded per person per unit time), and $t$ is occupation span.[2] From this simple equation, the number of artifacts discarded at a site would be the product of the discard rate, number of site occupants, and occupation span. If we have a 100 percent sample of three excavated sites, it would not be possible to rank the three sites in order of occupation span unless we knew a priori the number of occupants at each site. Furthermore, we would have to assume that the per capita rate of discard remained constant. Therefore, the number of artifacts in a site reflects the actual occupation span, but more so the product of occupation intensity and discard rate.

Occupation span is perhaps more elusive than occupation intensity, despite being a variable that more commonly interests archaeologists. Nonetheless, there is good reason to believe that unlike total artifact counts, horizontal artifact densities should relate more directly to actual occupation span. By horizontal artifact densities, I refer to the number of artifacts per unit area in the horizontal dimension, regardless of the depth of the deposit or excavation. A number of studies have found that site size increases as a power function of the number of site occupants (e.g., Narrol 1962; Whitelaw 1991; Wiessner 1974):

$$s = ap^b \qquad (3.4)$$

where $p$ is the number of site occupants, $a$ is the per capita space requirement, and $b$ is an exponential scalar. Because hunter-gatherer camps are typically circular or semicircular in form and have no significant vertical dimension (i.e., multilevel housing), theoretical and empirical considerations suggest the exponent $b$ is typically close to 2 (see appendix; Whitelaw 1991; Wiessner 1974). Substituting 2 for the exponent $b$ and combining equations 3.3 and 3.4 allows one to model artifact density ($d/s$):

$$\frac{d}{a} = \frac{prt}{ap^2} = \frac{rt}{ap} \qquad (3.5)$$

From this equation, it should be clear that even with constant occupation spans, per capita space requirements, and discard rates, sites with larger populations will exhibit lower densities than those with fewer occupants. Presumably in circular camp arrangements, larger sites will have larger areas of empty, unoccupied space. While this relationship describes recent forager camps, there is reason to question whether it would hold for archaeological sites. In theory, if archaeological localities were excavated randomly (i.e., pure random selection of excavation units within an artifact scatter), this relationship would also hold for archaeological cases.

With the exception of limited testing, archaeologists generally do not excavate large empty areas of sites. In fact, we generally try to avoid excavating the "empty space" and focus our efforts on areas of high artifact densities. If excavations dominantly occur in areas of significant artifact accumulation, then "occupied site area" ($s_o$) can be shown to conform to the simplified equation (see appendix):

$$s_o = pa \qquad (3.6)$$

and therefore artifact density is calculated as:

$$\frac{d}{s_o} = \frac{prt}{pa} = \frac{rt}{a} \qquad (3.7)$$

Notice that the number of site occupants cancels out of the equation and does not impact artifact densities. Instead, artifact densities are a product of the occupation span and discard rate, divided by the per capita requirements for space.

In the case of multiple spatially overlapping occupations, equations 3.3 and 3.7 would be multiplied by a constant representing the number of occupations. In this case, artifact counts and densities would represent the cumulative occupation intensity and cumulative occupation span, respectively. If we were interested in the mean or distribution of occupation spans represented by a mixed assemblage, using only artifact counts or densities, we would be out of luck.

Measures of mean per capita occupation span are more difficult to derive, but in some ways they are more useful because they should be independent of the number of occupations present. The degree to which mean per capita occupation span reflects mean occupation span for a site with mixed assemblages is dependent on the stability of residential groups. If group membership is unchanging, mean per capita occupation span should be identical to occupation span, but if membership is highly variable with individuals coming and going, these two values will diverge, as might the kinds of assemblages that are generated.

If the archaeological record is seen as the product of individual agents operating in time and space, and the behavioral phenomena we wish to study are the cumulative products of individuals, then per capita measures should be a more accurate reflection of the by-products of those behaviors. We could ask: What are the technological consequences of a site occupied for 30 days by 30 people in which group membership remains constant, or alternatively in which group membership is highly variable because individuals are very residentially mobile? If the frequency of residential mobility conditions certain aspects of technology for individuals, we would expect the two situations to have dramatically different technological signatures, despite having equal occupation spans and intensities.

## Developing Formal Models of Mean Per Capita Occupation Span

In this section, I begin to develop a formal model of artifact accumulation that allows the estimation of mean per capita occupation span for lithic assemblages. I consider the following measures to monitor per capita phenomena because I assume that toolkits (sets of lithic artifacts) are maintained by individuals. If toolkits are maintained at some aggregate social level (e.g., families), these measures would instead monitor the behavior of the individual social units that form the basis of the lithic economy. A reasonable point of departure is Schiffer's (1975a:840, 1987:53) discard equation:

$$d_t = \frac{S}{L} t \quad (3.8)$$

where $d_t$ is the number of a given type of artifacts discarded as a function of time, $S$ is the number of such artifacts maintained in systemic context, $L$ is the average use-life, and $t$ is the occupation span. This can be solved for occupation span:

$$t = \frac{L}{S} d_t \quad (3.9)$$

Assume a band of hunter-gatherers maintains 30 projectile points ($S$) at any given time, and each point has an average use-life of 15 days ($L$). From equation 3.6, two projectile points should enter the archaeological record per day ($F_d$). Therefore, if complete excavation produced 400 projectile points, then from equation 3.7, it could be inferred that 200 days were spent at the site. This, however, says nothing about the number of occupations represented, nor the mean length of occupation. A single 200-day occupation would produce the same result as 200 one-day, 100 two-day, or 20 ten-day occupations. If for sake of simplicity, a single occupation is assumed, we would extrapolate, from this site alone, that the settlement system was characterized by approximately 1.8 residential moves per year. This could be seen as an interesting "fact" of prehistoric behavior, and it could be very wrong.

If the discard equation is modified to estimate artifact ratios as a function of occupation span, occupation span becomes irrelevant. The following equations estimate the numbers ($N_1$, $N_2$) of two artifact types deposited in a site as a function of occupation span ($t$):

$$\text{Artifact type 1} \quad N_1 = \frac{S_1}{L_1} t \quad \text{Artifact type 2} \quad N_2 = \frac{S_2}{L_2} t$$

$$\text{Artifact type 1:2} \quad \frac{N_1}{N_2} = \frac{S_1 L_2 t}{S_2 L_1 t} \quad (3.10)$$

Notice that artifact ratios should remain constant regardless of occupation span because time cancels out of the equation (see Mills 1989:fig. 2 for a graphical representation of equation 3.10). In actuality, even if this equation is an accurate reflection of the system, because tool discard is a probabilistic process, these ratios will be variable, particularly for short-term occupations. However, the sum of multiple short-term occupations both in terms of artifact numbers and ratios should be equivalent to a comparable long-term occupation.

How can the discard equation be modified to address the issue of mean occupation span? If artifact discard rate is not modeled as a constant, but instead as a function of occupation span, then artifact ratios can be used to address the problem of mixed deposits. To take a simple example, assume the discard rate number of a particular class of artifacts ($F_d$) decreases as a linear function of occupation span ($t$):

$$F_d = 20 - \frac{t}{2} \qquad (3.11)$$

Notice that equation 3.5 permits negative discard rates, but for the moment I will ignore this problem. I use this equation only for its simplicity and heuristic value. In this case, on day 2, 19 of these items should be discarded, on day 20 of the occupation 10 items are discarded, and on day 40 the discard rate drops to zero. The total number of artifacts deposited in the site as a function of time is calculated as the integral of this equation:

$$N = \int_0^t \left(20 - \frac{t}{2}\right) = \frac{t^2}{4} + 20t \qquad (3.12)$$

While it is still possible to estimate the number of artifacts deposited for a single occupation of length $t$, estimating the cumulative duration of occupation represented by $N$ artifacts is not a simple matter because occupations of different individual length but equal cumulative lengths will produce different results. For example, 300 artifacts could imply a single twenty-day occupation or approximately four four-day occupations, a total of sixteen days. Similarly, during a single twenty-day occupation, 300 artifacts will be discarded, but during four five-day occupations (twenty cumulative days), 375 artifacts will be discarded.

Although variable rates of artifact accumulation confuse estimates of cumulative occupation span, they are ideal for estimating mean occupation duration in mixed assemblages. Assume the rate of discard for a second type of artifact increases as the occupation is extended, perhaps because it gradually replaces type 1 artifacts (above) in the toolkit. Assume the discard rate for type 2 artifacts ($F_{d2}$) is:

$$F_{d2} = \frac{t}{2} \qquad (3.13)$$

where $t$ is the length of occupation in days. While this function allows the discard rate to continually increase, in reality it should stabilize at some point in time. This function, again, was chosen for simplicity. The number of type 2 artifacts deposited as a function of time ($N_2$) would be calculated as:

$$N_2 = \int_0^t \frac{1}{2} t = \frac{t^2}{4} \qquad (3.14)$$

Figure 3.2a graphically depicts equations 3.12 and 3.14. Notice that the rate of discard of type 1 artifacts decreases, while the discard rate of type 2 artifacts increases with time. The ratio of type 1 to type 2 artifacts deposited through time would be:

$$\frac{N_1}{N_2} = \frac{20t - \frac{t^2}{4}}{\frac{t^2}{4}} = \frac{80}{t} - 1 \qquad (3.15)$$

This equation is depicted graphically in figure 3.2b. Notice that in short-term occupations, the $N_1:N_2$ ratio is high, but it declines as the occupation is lengthened. Therefore, using artifact discard rates that are responsive to occupation duration provides a means of distinguishing multiple short-term occupations from a single long-term occupation.[3] For example, a four-day occupation according to equation 3.13 should yield an $N_1:N_2$ ratio of 19, no matter how many four-day occupations are mixed in one assemblage, and thirty-day occupations will always yield an $N_1:N_2$ ratio of approximately 1.7. The ratios will remain the same for multiple occupations because equations 3.10 and 3.12 are modified only by the addition of a constant representing the number of occupations, and this constant cancels out when the ratio is calculated.

Of course, the key to applying this model to archaeological sites is identifying aspects of lithic technology whose discard rates should respond to occupation span. It will be difficult to estimate actual occupation spans

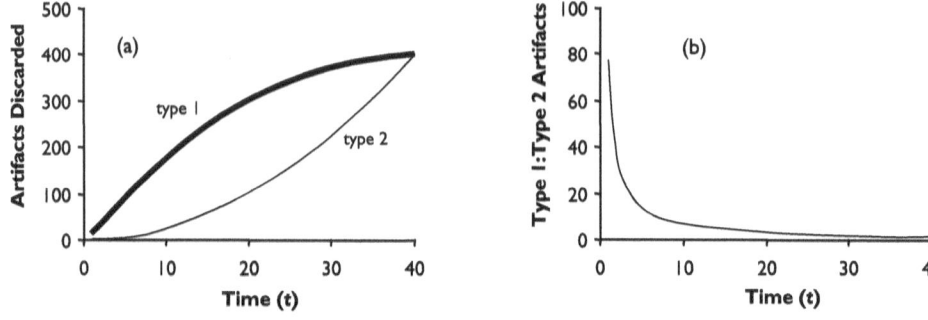

FIGURE 3.2. (a) Graph of equations 3.12 and 3.14 showing the cumulative number of type 1 and type 2 artifacts discarded as a function of time. (b) Graph of equation 3.15 showing the ratio of type 1 to type 2 artifacts as a function of time.

because use-lives and systemic numbers of artifacts are at best rough estimates and at worst unknowns, but comparing such measures across a range of sites allows their ordinal ranking.

## Modeling Mean Per Capita Occupation Span: Transported versus Locally Acquired Artifacts

The model developed in the prior section is problematic because, for one, it permits negative discard rates, and also because the constants I have chosen were never adequately explained, justified, or formalized. Equations 3.9 to 3.13 are solely intended to have heuristic value, a simplified form of the final model to be developed here.

This model assumes that a forager arrives at a site with a transported toolkit, what Schiffer (1975b:266) has called the "founding curate set." Upon arrival, the forager replenishes the toolkit to some optimum size of $k$ artifacts with locally acquired raw materials. The artifacts transported into the site are designated class a, and the artifacts acquired during the occupation of the site are designated class b. The model also assumes the following:

- $a_o$ = the number of artifacts of class a in a toolkit at time zero (the start of an occupation)
- $u$ = the mean use-life of an artifact

- $k$ = the optimal toolkit size
- Discard is probabilistic and is a function of use-life. The mean use-lives of transported and locally acquired artifacts are equal.
- Once artifacts are discarded, they are always replaced with class b artifacts.

With these assumptions, it is possible to calculate the number of class a artifacts in the toolkit as a function of time ($a_t$) as:

$$a_t = a_o e^{-\frac{t}{u}} \qquad (3.16)$$

This is the exponential growth/decay equation, perhaps most familiar to archaeologists as the radioactive decay equation with $1/u$ substituted for $\lambda$. In radioactive decay $\lambda$, the decay constant, is equal to the reciprocal of the mean life of a radioactive isotope. In this example, use-life is analogous to radioisotope mean life. Because discard is probabilistic, the likelihood of a class a artifact being discarded is proportional to the frequency of class a artifacts in the toolkit, and the radioactive decay equation correctly describes the system.

Archaeologically, however, we are not interested in the number of class a artifacts remaining in the toolkit, but instead the number discarded at a site. The cumulative number of class a artifacts discarded as a function of time ($d_{at}$) is calculated as:

$$d_{at} = a_o - a_o e^{-\frac{t}{u}} \qquad (3.17)$$

The number discarded ($d_{at}$) is the number of class a artifacts in the toolkit at time zero minus the number remaining in the toolkit at time $t$. This model does not allow negative discard rates, because the total number of artifacts discarded is limited by the number transported into the site. The cumulative number of class b artifacts discarded as a function of time ($d_{bt}$) is calculated as:

$$d_{bt} = \frac{kt}{u} - \left(a_o - a_o e^{-\frac{t}{u}}\right) \qquad (3.18)$$

where $kt/u$ represents the total number of artifacts discarded for time $t$. In figure 3.3, these equations are depicted graphically. The rate of discard of class a artifacts decreases as occupation is lengthened, as they become increasingly rare in the toolkit. Eventually, all class a artifacts are discarded, and the graph approaches an asymptote at $d_{at} = a_o$. As class b

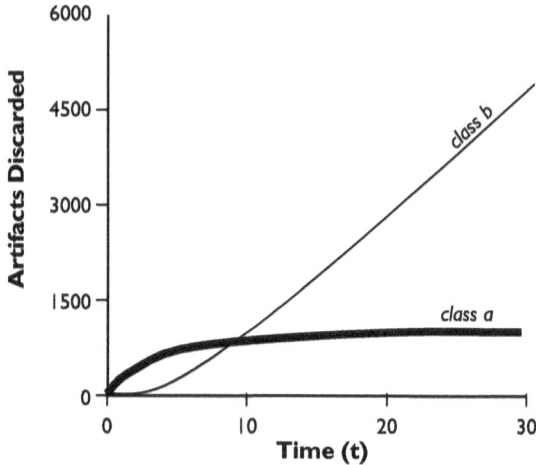

FIGURE 3.3. Graph of equations 3.17 and 3.18, the cumulative number of class a and class b artifacts discarded as a function of time, assuming $k=1{,}000$, $a_o=1{,}000$, $u=5$.

artifacts replace class a artifacts in the toolkit, they become more common, and the discard rate increases until all class a artifacts have been discarded. At this point, the discard rate for class b artifacts stabilizes at $k/u$.

From figure 3.3, it is evident that the ratio of class b to class a artifacts should provide a measure of mean occupation span since discard rates do change as a function of time. The ratio is calculated as:

$$\frac{d_{bt}}{d_{at}} = \frac{\frac{kt}{u} - \left(a_o - a_o e^{-\frac{t}{u}}\right)}{a_o - a_o e^{-\frac{t}{u}}}$$

$$\frac{d_{bt}}{d_{at}} = \frac{kt}{u\left(a_o - a_o e^{-\frac{t}{u}}\right)} - 1 \qquad (3.19)$$

Equation 3.19 is depicted graphically in figure 3.4a, and its reciprocal is shown in figure 3.4b. These graphs show that class b artifacts become increasingly abundant as occupation span increases. In this example the ratio of class b to class a artifacts increases in a nearly linear fashion. Notice, however, that the reciprocal ratio (a:b) takes a very different form, where very short-term occupations are dominated by class a artifacts, and the ratio drops quickly as occupation duration is lengthened. Therefore, the former ratio (class b:class a) should be a much more reliable indicator of

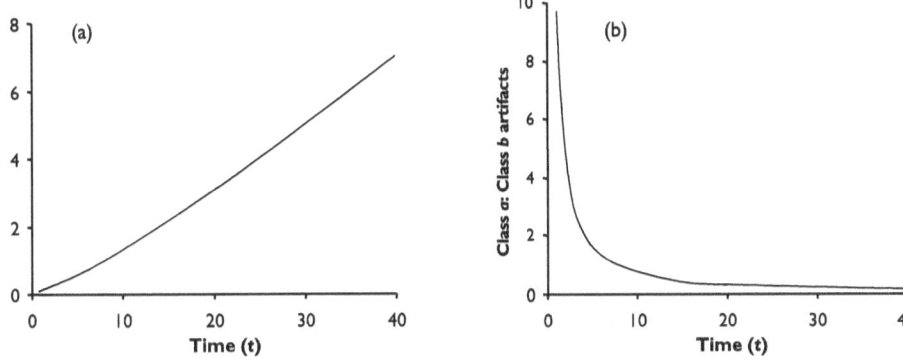

FIGURE 3.4. (a) Graph of equation 3.19, the ratio of class b:class a artifacts vs. time, assuming $k = 1,000$, $a_o = 1,000$, $u = 5$. (b) Graph of the ratio of class a:class b artifacts vs. time, the reciprocal of equation 3.19.

mean occupation span than the latter (class a:class b) unless one wants to look only at very coarse differences.

## A Closer Look at the Model

The model described in the previous section is a very simple idea, expressed in a somewhat complex way. The model has the rather unsurprising and unoriginal prediction that *as occupation span is lengthened, archaeological assemblages will become increasingly dominated by artifacts acquired locally.* Short-term occupations should have relatively high proportions of transported artifacts, because in the early phases of an occupation things brought into the site from elsewhere will dominate toolkits. Long-term occupations should have relatively high proportions of local artifacts, because in the later stages of an occupation things acquired locally will dominate toolkits. This ratio should monitor mean per capita occupation span, not cumulative occupation span. Also, the ratio of local: transported (class b:class a) artifacts should be a more precise measure of mean per capita occupation span than its reciprocal.

As reasonable as this may sound, applying this generalization to a sample of archaeological sites requires the addition of the mantra of modelers everywhere, "all things being equal." An examination of equation 3.19 shows why. The ratio of local:transported artifacts is not solely dependent on time. The size of the toolkit ($k$), the state of toolkit upon arrival ($a_o$), and the mean use-life of artifacts ($u$) will all impact this measure of occupation span. The effects of these variables are shown in figure 3.5. As the initial state of the transported toolkit becomes increasingly depleted, the ratio of locally acquired to transported artifacts is increased while holding time constant (fig. 3.5a). The opposite effect is seen with larger toolkit sizes (fig. 3.5b). Additionally, as mean artifact use-lives increase, the rate of change in the ratio over time is decreased, although it remains constant at the start of an occupation (fig. 3.5c). The ratio could also be affected if the mean use-lives of artifacts in the transported and local toolkits differ. Since artifacts in transported toolkits are often designed to have long use-lives (Bamforth 1986; Larson 1994; Shott 1989), this problem may be common. Therefore, depending on the initial starting conditions and state of the technological system at individual sites, the ratio of local to transported artifacts may begin at differing values and follow different trajectories through time. And more practically, I assume that the transported and local components of an assemblage can be identified. Nonetheless, the original prediction, that locally acquired artifacts should increasingly dominate sites as occupations are prolonged, should always hold true, but the knowledge of these potentially complicating factors necessitates its cautious use.

## Application: The Study Sample

To apply the model, it is necessary to differentiate between locally acquired and transported artifacts. Lithic raw material is likely the most straightforward correlate of artifact transport status. Local raw materials are assumed to have been acquired during an occupation, and nonlocal raw materials are assumed to have been acquired prior to the occupation. I assume that raw materials exceeding a 20 km linear distance from a site are nonlocal. This distance is intended to represent the maximum distance a pedestrian forager could reasonably traverse in one day. In other words, it is the distance at which the probability of the acquisition of materials during the occupation of a site becomes extremely low, such that materials acquired from beyond 20 km can be assumed with some confidence to have been acquired prior to the start of an occupation. Observations of recent hunter-gatherers would suggest that 20 km one-way foraging distances are extremely rare (Binford 2001:235–238).

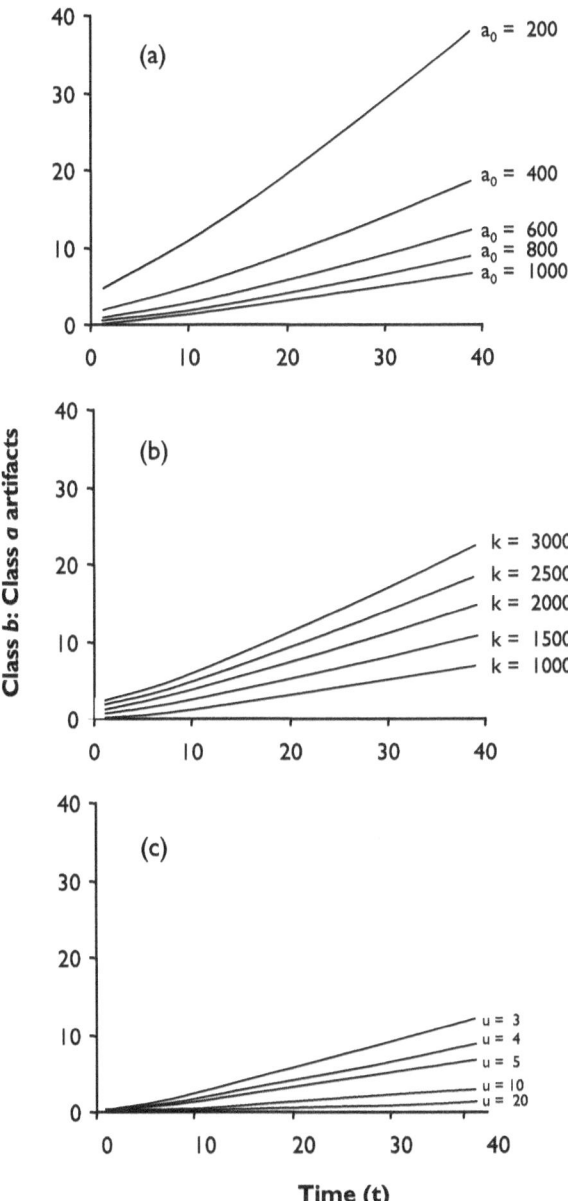

FIGURE 3.5. The effects of (a) the size of the initial transported toolkit ($a_o$), (b) the optimal size of the toolkit ($k$), and (c) the mean artifact use-life ($u$) on the ratio of locally acquired (class a) to transported (class b) artifacts as a function of time, assuming $k = 1{,}000$, $a_o = 1{,}000$, $u = 5$, except as shown for each respective graph.

Three problems are inherent to applications of the model. First, it can be applied only to assemblages composed of both local and nonlocal raw materials. If one is lacking, the assumptions of the model are violated because, in effect, there would be either no transported or no local toolkit represented. Second, the possibility exists that some proportion of raw materials treated as local were acquired prior to a site occupation. For example, oolitic and algal cherts are abundant in the vicinity of the Krmpotich site and are well represented in the lithic assemblage. However, visually indistinguishable raw materials are known to outcrop in other more distant areas of southwest Wyoming, such as on Delaney Rim, roughly 110 km to the southeast (Love 1997). Therefore, it is possible and perhaps likely that some of the oolitic and algalitic chert present at Krmpotich may have been imported from a previous campsite. Similarly, it is possible that a portion of the nonlocal raw materials considered transported was acquired during occupation. Acknowledging this, all raw materials that are locally available, for the purposes of this exercise, are treated as locally acquired, and all raw materials not locally available are treated as transported into the site from a previous residential occupation.

For the five sites in the study sample, artifact assemblages were tabulated according to artifact type and raw material. For each artifact class, lithic raw materials were classified as local, nonlocal, or unknown. To avoid sampling problems associated with different excavation and screening practices among the sites, only artifacts larger than 2 cm were included in the analysis. Table 3.1 shows the breakdown of artifact counts by raw material and artifact type and the ratio of local:nonlocal raw materials for each site.

The ratios are highly variable, ranging from 0.09 for Carter/Kerr-McGee (CKM) to 240.14 for Barger Gulch, Locality B (BGB). Superficially, the ratios express a large range of mean per capita occupation spans for these sites. Of particular note is a ratio of 5.00 for Upper Twin Mountain, a briefly occupied bison kill site, which exceeds the ratios of both Carter/Kerr-McGee and Agate Basin campsites.

There is, however, good reason to believe that the assemblage from Twin Mountain suffers from some of the problems discussed above. For example, although much of the assemblage is composed of locally available Troublesome Formation chert, including one projectile point (a surface

TABLE 3.1. Artifact raw material frequencies, type frequencies, and local to nonlocal raw material ratios by site for all artifacts larger than 2 cm in maximum dimension.

| Artifact type–Raw material | Site | | | | |
|---|---|---|---|---|---|
| | CKM | AGB | KRMP | BGB | UTM |
| Debitage–Local | 2 | 182 | 473 | 1,503 | 5 |
| Debitage–Nonlocal | 15 | 187 | 19 | 0 | 0 |
| Debitage–Unknown | 0 | 14 | 8 | 0 | 0 |
| Tools–Local | 0 | 10 | 36 | 113 | 4 |
| Tools–Nonlocal | 12 | 56 | 15 | 3 | 0 |
| Tools–Unknown | 2 | 7 | 5 | 0 | 0 |
| Channels–Local | 0 | 4 | 7 | 9 | 0 |
| Channels–Nonlocal | 5 | 39 | 2 | 2 | 0 |
| Channels–Unknown | 0 | 3 | 0 | 0 | 0 |
| Cores–Local | 1 | 5[†] | 3 | 31 | 1 |
| Cores–Nonlocal | 0 | 0 | 0 | 0 | 0 |
| Cores–Unknown | 0 | 0 | 1 | 0 | 0 |
| Bifaces–Local | 0 | 6 | 9 | 13 | 0 |
| Bifaces–Nonlocal | 0 | 1 | 0 | 1 | 0 |
| Bifaces–Unknown | 0 | 0 | 1 | 0 | 0 |
| Points–Local | 0 | 0 | 4 | 12 | 0 |
| Points–Nonlocal | 3 | 15 | 1 | 1 | 2 |
| Points–Unknown | 0 | 3 | 0 | 0 | 0 |
| Sum | 40 | 532 | 584 | 1,688 | 12 |
| Sum local | 3 | 207 | 532 | 1,681 | 10 |
| Sum nonlocal | 35 | 298 | 37 | 7 | 2 |
| Local:nonlocal | 0.09 | 0.69 | 14.38 | 240.14 | 5.00 |

*Notes:* Site abbreviations: CKM = Carter/Kerr-McGee, AGB = Agate Basin, Area 2, KRMP = Krmpotich, BGB = Barger Gulch, Locality B, UTM = Upper Twin Mountain. [†]Although only three cores were located for analysis, Frison (1982b:70) notes that five were recovered from the component.

find), it is reasonable to question whether this raw material was imported into the site before the occurrence of the kill. Also, because the Twin Mountain assemblage represents a relatively short-term occupation, as discussed above, artifact representation is potentially biased by the probabilistic nature of discard. Nonetheless, the large range of values for this ratio hints at extreme variation in Paleoindian mean per capita occupation spans.

## A Second Model: Debitage versus Transported Tools

To test the first model, an independent measure of mean per capita occupation span is needed. This model is similar to the previous one, except the definition of local and transported artifacts is modified. The application of the previous model assumed that a transported artifact is one whose raw material was acquired prior to the occupation of a site. The definition in the second model instead hinges on where an artifact was manufactured, regardless of when its raw material was acquired. Assume the following:

- Debitage is not transported. All debitage is produced locally.
- Tools are transported into sites and produced locally. Tools are defined as all objective pieces (Andrefsky 1998:10–11), including flake tools, cores, bifaces, and projectile points.
- Discard of tools is probabilistic. All tools in the toolkit have an equal probability of discard.
- $u_y$ = the use-life of tools.
- $u_f$ = the use-life of debitage.
- $a_o$ = the number of tools transported into the site.
- $k_f$ = the quantity of debitage maintained in the toolkit. $k_f$ refers not only to debitage itself, but also the potential to produce debitage (i.e., all objective pieces).

Using these assumptions, it is possible to apply the second model to sites lacking local raw materials, as is commonly the case with Paleoindian sites. Again however, this model cannot be applied to sites lacking nonlocal raw materials. The number of transported tools discarded as a function of time ($d_t$) is calculated as:

$$d_t = a_o - a_o e^{-\frac{t}{u_y}} \qquad (3.20)$$

The quantity of debitage discarded ($d_f$) as a function of time is:

$$d_f = \frac{k_f t}{u_f} \qquad (3.21)$$

The derivation of equation 3.21 follows that of Schiffer's discard equation. Notice that, unlike in the previous model, the quantity of debitage discarded is a linear function of time since it is assumed that all debitage is produced locally. Although the discard rate of debitage does not change as a function of time, the discard rate of nonlocal tools is affected by time, so their ratio can provide a measure of mean occupation span:

$$\frac{d_f}{d_t} = \frac{\overline{\frac{k_f t}{u_f}}}{a_o - a_o e^{-\frac{t}{u_y}}} = \frac{k_f t}{u_f a_o \left(1 - e^{-\frac{t}{u_y}}\right)} \quad (3.22)$$

Graphical representations of equation 3.22 and its reciprocal are depicted in figure 3.6. Through time, assemblages become increasingly dominated by debitage relative to tools transported into the site. The application of this model for the estimation of mean per capita occupation span also depends on assuming "all things being equal." In particular, if use-lives or inventories of debitage and transported tools vary between sites, the measure will respond to other factors in addition to mean per capita occupation span.

## An Application and a Test

Before I and you the reader applaud ourselves for slogging through these equations and deriving a quantifiable measure of mean per capita occupation span, we must step aside and ask the thousand-dollar question: do these variables work? If both variables monitor mean per capita occupation span, the ratios of local:nonlocal artifacts and debitage:nonlocal tools should be correlated for a set of archaeological sites. Unfortunately, this test is complicated by autocorrelation since the term "transported tools" appears in the denominators of both models. In figure 3.7, I have created 500 random archaeological assemblages of 10,000 artifacts, comprising local debitage, local tools, transported debitage, and transported tools. Using these four categories of artifacts, it is possible to explore the codependence of the two models. The ratios are very weakly positively correlated ($r = 0.167$; $r^2 = 0.28$; $p < 0.001$). The correlation is stronger when both ratios are logged, but the correlation remains quite weak ($r = 0.370$; $r^2 = 0.137$; $p < 0.001$). Still, using the models as independent tests of each other is statistically not as straightforward as it would be if they were entirely independent measures.

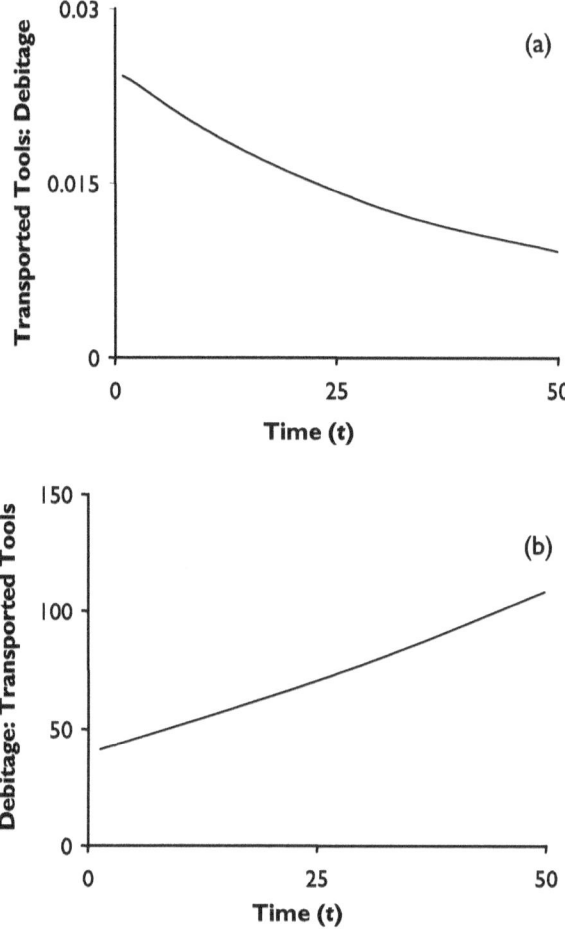

FIGURE 3.6. (a) A graphical representation of equation 3.22 showing the relationship between time and the ratio of debitage:transported tools, and (b) its reciprocal. Both graphs assume $a_0 = 50$, $k_f = 1,000$, $u_y = 20$, $u_f = 10$.

Figure 3.8 shows the relationship between the logged and unlogged ratios of local:nonlocal raw material and debitage:nonlocal tools for the five sites in the study sample (table 3.2). For the unlogged ratios (fig. 3.8a), Barger Gulch stands out as an outlier being dominated by both local raw materials and debitage. The remaining sites form a tight cluster around the origin, possibly suggesting a bimodal distribution in mean per capita occupation spans. The ratios are highly correlated ($r = 0.999$; $p < 0.001$), but this is largely due to the shape of the point scatter. When

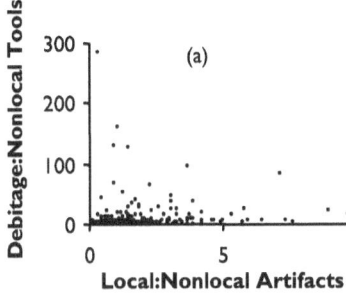 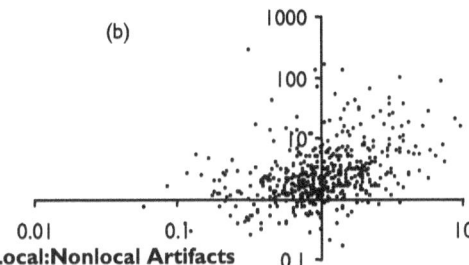

FIGURE 3.7. (a) Scatterplot of local:nonlocal raw materials and debitage: nonlocal tools for 500 random archaeological assemblages of 10,000 artifacts. Artifacts were randomly distributed into local debitage, local tools, nonlocal debitage, and nonlocal tools. (b) Same with both axes logged (Pearson's r = 0.4, $r^2$ = 0.16).

TABLE 3.2. Raw material totals and debitage and nonlocal tool counts tabulated by site for all artifacts larger than 2 cm.

| Variable | Site | | | | |
|---|---|---|---|---|---|
| | CKM | AGB | KRMP | BGB | UTM |
| Local artifacts | 3 | 207 | 532 | 1,681 | 10 |
| Nonlocal artifacts | 35 | 298 | 37 | 7 | 2 |
| Nonlocal tools | 15 | 72 | 16 | 5 | 2 |
| Debitage | 22 | 429 | 509 | 1,514 | 5 |
| Local:nonlocal | 0.09 | 0.69 | 14.38 | 240.14 | 5.00 |
| Log (local:nonlocal) | −1.07 | −0.16 | 1.16 | 1.38 | 0.70 |
| Debitage:nonlocal tools | 1.47 | 5.96 | 31.81 | 302.8 | 2.50 |
| Log (debitage:nonlocal tools) | 0.17 | 0.78 | 1.50 | 2.48 | 0.40 |

the ratios are logged (fig. 3.8b), a strong correlation is still evident (r = 0.898; p = 0.038), suggesting that the two ratios are monitoring the same phenomenon, namely mean per capita occupation span. Because of the autocorrelation problem, however, Pearson's product moment correlations cannot reliably assign probabilities to these point scatters.

A Monte Carlo routine was used to evaluate the significance of the correlation of the logged ratios. Ten thousand sets of random archaeological assemblages were created for five sites using the number of artifacts larger than 2 cm observed in each of the five sites in the study sample.

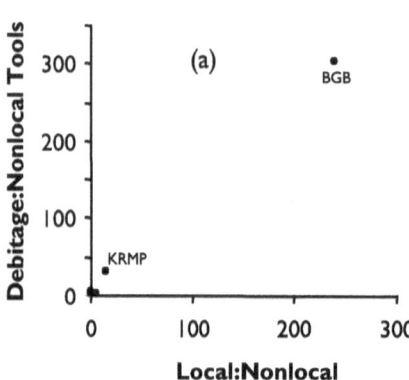 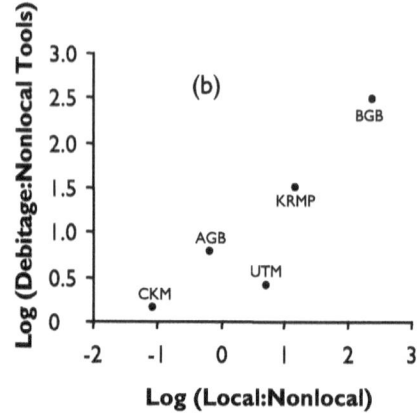

FIGURE 3.8. (a) Local:nonlocal raw materials vs. debitage:nonlocal tools for the five sites of the study sample. (b) Same with both ratios logged.

Again, artifacts were randomly distributed between local and nonlocal tools and debitage. For each random set of sites, Spearman's ρ was calculated.[+] The p-value for the observed distribution of ratios was calculated as the proportion of random assemblages exceeding the observed nonparametric correlation coefficient (ρ = 0.90). The Monte Carlo estimated p-value for the observed correlation of ratios among the five sites is not significant (one-tailed p = 0.125).

Interestingly, the Upper Twin Mountain site, the only bison kill in the sample, appears to break the trend. In particular, its rank falls below that of Agate Basin, with respect to the debitage:nonlocal tool ratio, but it again stands out with respect to the local:nonlocal raw material ratio (fig. 3.8b). Because the site differs from the remainder of the sites in function, there may be good reason to expect it to break from the trend. Since we can reasonably expect most of the lithic raw material present at bison kill sites to be transported there from a nearby camp (because it seems unlikely that hunters encountered prey, fashioned weapons, and made the kill simultaneously), the raw material signature of Upper Twin Mountain may reflect the occupation span of the campsite from which the hunting party originated rather than the site itself. Furthermore, the site has a very small sample of artifacts larger than 2 cm (n = 12), which may also contribute to the problem. However, when the Upper Twin Mountain site is removed, and the Monte Carlo routine is repeated, the correlation for the four campsites is still not significant (Pearson's r = 0.997; Spearman's ρ = 1.000; one-tailed p = 0.108), despite a perfect positive nonparametric correlation. Because no greater correlation is possible, it is simply impossible to achieve a significant correlation using this statistical method with a sample of only four sites.

Therefore, an increased sample of Folsom and Goshen camp and kill sites was used. Because I had to rely on published data, only an additional six sites could be added to the sample: Hanson (Ingbar 1992), Lake Theo (Buchanan 2002), Cooper's Middle and Lower Kills (Bement 1999), Mill Iron (Francis and Larson 1996), and Bobtail Wolf (Root et al. 2000). Table 3.3 shows the ratios of local:nonlocal raw materials and debitage:nonlocal tools for the enlarged sample. Also, for comparative purposes, I have included the recalculated ratios for the five sites in the study sample using the entire sample of excavated artifacts from the sites. A comparison of tables 3.2 and 3.3 shows how the ratios in the study sample sites are affected by the inclusion of all excavated artifacts. In general, local:nonlocal ratios decline when all artifacts are used, and debitage:nonlocal tool ratios increase. For example, at Barger Gulch the ratio of local:nonlocal raw materials decreases to 88.23 for all artifacts from 240.14 for artifacts larger than 2 cm. Also, the debitage:nonlocal tools ratio increases from 302.8 for artifacts larger than 2 cm to 2,774.14 for all artifacts. The direction of these changes is constant for all sites, showing that smaller artifacts in assemblages tend to be characterized by increased proportions of nonlocal raw materials and debitage. The magnitude of these changes, however, does vary between sites and highlights the difficulty of comparing data in the study sample with those from published sources where size control is not available. Nonetheless, the patterning seen in the enlarged sample is compelling (fig. 3.9).

Figure 3.9a shows the relationship between the unlogged ratios. All but three sites cluster around the origin. Hanson, Barger Gulch, and Bobtail Wolf deviate from the cluster, suggesting relatively long-term occupations. Bobtail Wolf in particular exhibits especially high ratios, suggesting an extremely prolonged occupation, but this may be in part an artifact of sampling, as discussed above. Also, all of the kill sites reasonably appear as relatively short-term occupations, tightly hugging the origin in the graph. More convincing, however, is figure 3.9b, showing the relationship between the logged ratios. The enlarged set of eight campsites and three kill sites reinforces the patterning seen in the study

TABLE 3.3. Ratios of local to nonlocal raw material and debitage to nonlocal tools for all artifacts in the study sample and from published data in the enlarged sample.

| Site | Site Type | n | L:N | D:NT | Log (L:N) | Log (D:NT) | Citation |
|---|---|---|---|---|---|---|---|
| Study sample (all artifacts) | | | | | | | |
| Carter/Kerr-McGee | Camp | 1,990 | 0.005 | 98.25 | −2.34 | 1.99 | |
| Agate Basin, Area 2 | Camp | 17,920 | 0.16 | 214.35 | −0.80 | 2.33 | |
| Krmpotich | Camp | 8,866 | 11.33 | 461.53 | 1.05 | 2.66 | |
| Barger Gulch, Loc. B | Camp | 19,638 | 88.23 | 2,774.14 | 1.95 | 3.44 | |
| Upper Twin Mountain | Kill | 87 | 1.42 | 39.50 | 0.15 | 1.60 | |
| Enlarged sample | | | | | | | |
| Bobtail Wolf | Camp | 394,473 | 645.68 | 3,418.15 | 2.81 | 3.53 | Root et al. 2000 |
| Cooper Lower Kill | Kill | 26 | 0.07 | 0.60 | −1.18 | −0.22 | Bement 1999:52–124 |
| Cooper Middle Kill | Kill | 55 | 0.02 | 5.75 | −1.73 | 0.76 | Bement 1999:52–124 |
| Hanson[†] | Camp | 5,087 | 120.12 | 259.39 | 2.08 | 2.41 | Ingbar 1992 |
| Lake Theo | Camp | 173 | 2.76 | 5.39 | 0.44 | 0.73 | Buchanan 2002 |
| Mill Iron | Camp/Kill | 1,710 | 21.41 | 87.39 | 1.33 | 1.94 | Francis and Larson 1996 |

*Notes:* [†]Values for Hanson assume Phosphoria Formation chert is a local raw material. Abbreviations: n = sample size, L:N = local:nonlocal raw materials, D:NT = debitage:nonlocal tools.

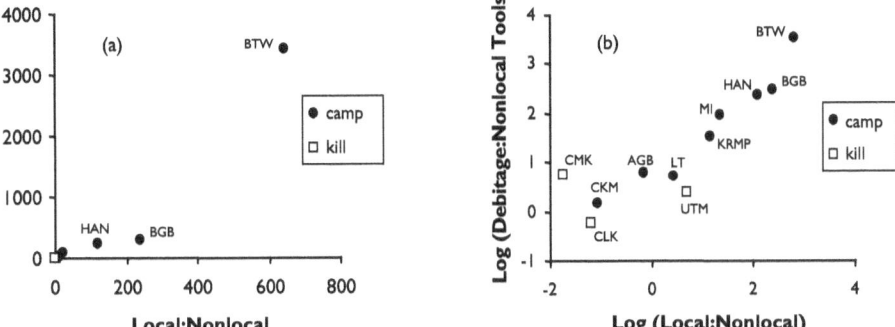

FIGURE 3.9. (a) Scatterplot of local:nonlocal raw materials vs. debitage:nonlocal tools for the enlarged sample of sites. (b) Same with logged values. Abbreviations: AGB = Agate Basin; BGB = Barger Gulch, Loc. B; BTW = Bobtail Wolf; CKM = Carter/Kerr-McGee; CLK = Cooper Lower Kill; CMK = Cooper Upper Kill; HAN = Hanson; LT = Lake Theo; MI = Mill Iron; KRMP = Krmpotich; UTM = Upper Twin Mountain. Study sample sites plotted for all artifacts larger than 2 cm.

sample alone (fig. 3.8). Although Mill Iron contains both a camp and a kill area, I included it within the campsites because the majority of the artifacts are derived from the residential area. Within campsites, the logged ratios of local:nonlocal raw materials and debitage:nonlocal tools are highly correlated (Pearson's r = 0.965; Spearman's ρ = 0.827). The Monte Carlo routine was again repeated using eight sites with the observed sample sizes (table 3.3), and the correlation was found to be significant (p = 0.0247). Thus, both ratios are likely monitoring the same underlying variable. Finally, as in the study sample, the kill sites deviate from the pattern seen in campsites, possibly an indication that kill sites violate the assumptions of the models, or possibly an indicator of small sample size.

Returning to the question of the Hanson site, this analysis might support Frison and Bradley's interpretation of a fairly long-term occupation, grouping with Barger Gulch and Bobtail Wolf. The placement of Hanson, however, is dependent on whether Phosphoria chert is treated as a local or nonlocal raw material. In this exercise, I have included Phosphoria in the sample of local raw materials since Frison and Bradley place the source area 19.8 km from the site and Ingbar places it at 20 km. These values fall right on the cusp of my distinction between local and

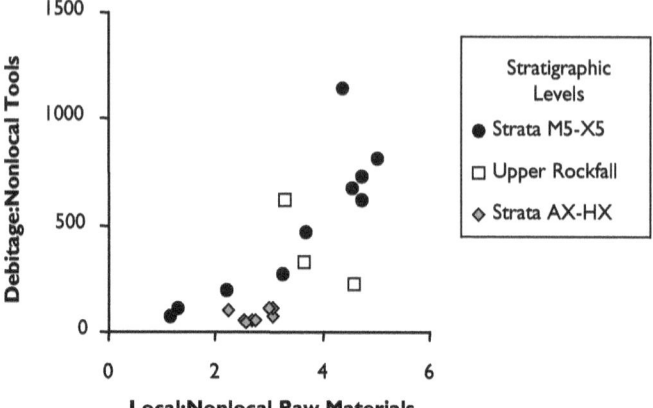

FIGURE 3.10. Local:nonlocal raw materials vs. debitage:nonlocal tools for twenty-one stratigraphic layers of Puntutjarpa rockshelter, Australia. Markers by stratigraphic layer. Puntutjarpa data are from Gould (1977).

nonlocal raw materials. Interestingly, if Phosphoria is instead treated as a nonlocal raw material, Hanson remains on the trend line, but instead falls in with Mill Iron and Krmpotich (local:nonlocal = 11.43; debitage:nonlocal tools = 48.63), suggesting a mean occupation length that falls within an intermediate range for Folsom and Goshen sites, more in line with Ingbar's (1992) interpretation. The Hanson question will have to remain unresolved, at least for this chapter.

The tests support the hypothesis that both measures are monitoring the same phenomenon, which is argued to be mean per capita occupation span. Kill sites, commonly dominated by or exclusively composed of transported artifacts, for good reason, do not fit the trend. Kill sites do, however, consistently appear to be short-term occupations. In part, this is a product of small sample size because high ratios cannot be produced from small assemblages. For example, an assemblage with only four artifacts can only produce ratios between 0 and 3. This does not mean that the method cannot be applied to sites with small assemblages. After all, small assemblages should indicate brief occupations, and in that light, the method is monitoring exactly what it is supposed to monitor, but it does warrant careful use of the method with sites where excavations are limited in scope. The more important application is for large sites with large assemblages because it is in these sites that the question of reoccupation is more of an issue.

## Still Not Convinced?

"Correlation is not causation" is an oft-repeated saying in statistics courses. This phrase applies well to the correlations performed in the previous section because it was not argued that the ratio of local:nonlocal raw materials "causes" change in the ratio of the debitage:nonlocal tools or vice versa. Instead, it is assumed, based on a theoretical formal model, that the underlying causal (or confounding) variable driving both ratios is mean per capita occupation span. However, without really knowing what site occupation spans were, the only reason to believe that these ratios monitor what I claim them to is that the models I have built say that they should. In fact, it is possible that the observed variation is driven by some other variable, such as discard rate. But before you bemoan having read the previous equation-filled pages, it is possible to independently test my claims. An ideal test case would involve a set of sites, perhaps from an ethnoarchaeological context, where mean per capita occupation and ratios of transported to local artifacts are known. I know of no such available data set. Another approach would be to compare measures of mean per capita occupation span to an independent variable that is known to affect mobility.

It is well known that various aspects of hunter-gatherer mobility, when examined globally, correlate well with environmental variables (Binford 1980, 2001; Kelly 1983, 1995). The marginal value theorem predicts that the timing of residential moves should be governed by patch-specific (site-specific) gain functions, average environmental returns, and distances between patches, measured with respect to some currency (Charnov 1976b; Kelly 1995). To use this approach with the Paleoindian sites of the study sample would require independent environmental data for variables that might govern mobility, such as season of occupation, prey return rates, average environmental return rates, distances between possible camping locations, and so on, and these data are woefully incomplete or unknown. Instead, I turn to Punututjarpa rockshelter from the Western Desert of Australia (Gould 1977). That the availability and distribution of water strongly affects hunter-gatherer mobility in arid regions is well known (Gould 1969, 1977; Lee 1979; Yellen 1977), and paleoenvironmental data from Australia provide one independent data set for comparison with archaeological data from Punututjarpa.

*Puntutjarpa Rockshelter*

Richard Gould's (1977) excavation of Puntutjarpa rockshelter in the Western Desert of Australia from 1969 to 1970 provides an excellent test case. This site contains approximately 10,000 $^{14}$C years of continuous deposition, with evidence of human occupation throughout. Although three areas of the rockshelter were excavated, this analysis will focus on the "Trench 2" excavations because the best tabulated data are available from this portion of the site. Both local and nonlocal raw materials are present, and their source areas are known (Gould 1977:123–127). The site was excavated in arbitrary three- and six-inch levels within natural strata. Gould's assemblage data are tabulated for twenty-three arbitrary stratigraphic levels. These arbitrary stratigraphic divisions are not believed to equate to twenty-three individual occupations.

Puntutjarpa was likely repeatedly occupied because a small overhang in a quartzite outcrop provided shelter. Though Gould (1977:76–79) was somewhat puzzled by the lack of available water at the site, the occurrence of a single rainstorm in December of 1969 provided hints to potential water sources. After this storm, the sandy sediments of the West Cave of the shelter retained moisture for at least a month, and one excavated feature (Feature 44) from the West Cave was interpreted as a "native well." Also, the rainfall activated an ephemeral spring, which seeped from a stream bank 500 m to the northwest for more than one week. If regional water availability and local spring discharge governed lengths of stay at Puntutjarpa, then a direct relationship between aridity and mobility is expected. Whether greater aridity would lead to shorter or longer occupations, however, is debatable. Arguing from the marginal value theorem, greater aridity would mean less spring discharge, and therefore water returns would reach zero quickly. In other words, if people stayed until water ran out, then decreased spring discharge would translate to shorter occupations. On the other hand, if decreased regional water availability led to greater distances between suitable camping sites, then the distances between "patches" would be increased, leading to longer occupations. Either way, a direct correlation, whether positive or negative, is expected between the availability of water and the length of occupations.

Lithic raw materials were classified as local and nonlocal, using the 20 km cutoff distance. Nonlocal raw materials from Puntutjarpa include

TABLE 3.4. Artifact density, local:nonlocal raw materials, debitage:nonlocal tools, and occupation span index by stratigraphic level for Puntutjarpa rockshelter.

| Stratum | Excavation area (m²) | Artifacts | Art. density (n/m²) | Local: nonlocal | Debitage: nonlocal tools | Occupation span index |
|---|---|---|---|---|---|---|
| AX | 15.88 | 770 | 48.5 | 2.25 | 107.57 | 27.15 |
| BX | 15.26 | 1,537 | 100.7 | 3.09 | 75.71 | 34.14 |
| CX | 14.64 | 2,622 | 179.1 | 2.67 | 56.22 | 29.14 |
| DX | 14.02 | 3,806 | 271.5 | 2.76 | 58.16 | 30.14 |
| EX | 13.70 | 4,535 | 331 | 2.52 | 58.23 | 27.69 |
| FX | 13.69 | 3,727 | 272.2 | 2.57 | 50.55 | 27.83 |
| GX | 13.68 | 4,792 | 350.3 | 3.06 | 110.79 | 35.36 |
| HX | 13.67 | 3,687 | 269.7 | 3.01 | 112.84 | 34.95 |
| IZ | 10.03 | 3,116 | 310.6 | 3.30 | 615.60 | 59.82 |
| JZ | 7.53 | 1,335 | 177.4 | 3.67 | 328.75 | 50.96 |
| KZ | 5.02 | 1,606 | 320.1 | 4.60 | 227.29 | 55.8 |
| LZ | 2.51 | 636 | 253.5 | 4.58 | | |
| M5 | 13.82 | 8,146 | 589.4 | 4.72 | 732.82 | 79.08 |
| N5 | 13.79 | 9,845 | 714.1 | 5.01 | 813.67 | 85.46 |
| O5 | 13.75 | 8,768 | 637.5 | 4.72 | 620.79 | 74.22 |
| P5 | 13.72 | 6,140 | 447.5 | 4.55 | 676.22 | 74.93 |
| Q5 | 13.65 | 4,657 | 341.1 | 4.38 | 1,147.25 | 93.78 |
| R5 | 13.55 | 3,790 | 279.6 | 3.70 | 468.13 | 57.32 |
| S5 | 13.46 | 1,908 | 141.8 | 3.24 | 268.29 | 44.05 |
| T5 | 13.36 | 1,411 | 105.6 | 2.21 | 195.57 | 30.64 |
| U5 | 10.63 | 456 | 42.9 | 1.29 | 108.50 | 17.63 |
| V5 | 5.24 | 76 | 14.5 | 1.17 | 71.00 | 14.79 |
| X5 | 2.73 | 18 | 6.6 | 0.38 | | |

*Note*: Shaded rows denote levels containing the Upper Rockfall. Data from Gould (1977:tables 25–33, figs. 33, 34, 66, 67). See text for explanation of the occupation span index.

white chert, the closest source of which is 22.4 km to the north, and a group of raw materials Gould (1977:123–127) refers to as "exotic," defined as having source areas greater than 40 km (25 mi) from the site. Gould (1977:tables 25–33, figs. 66–67) also provides counts of debitage, tools, and cores broken down by lithic raw material, allowing the calculation of the debitage:nonlocal tool ratio. The local:nonlocal and debitage: nonlocal tool ratios for the twenty-three arbitrary levels are shown in table 3.4. Two levels (LZ and X5) lack tools manufactured on nonlocal raw materials, making the debitage:nonlocal tool ratio impossible to calculate. Figure 3.10 shows the correlation between the two ratios. Standard Pearson's and Spearman's correlations show significant positive correlations (r = 0.757; ρ = 0.799; p < 0.001). Based on 10,000 randomized assemblages in the Monte Carlo simulation, the observed Spearman's ρ cannot be attributed to autocorrelation (p = 0.003). Though the two ratios are correlated for the entire assemblage, the upper and lower levels of the shelter do appear to be behaving differently (fig. 3.10). This may be due to changes in site function following a major rockfall event, referred to as the Upper Rockfall (Gould 1977:176–177).

Though paleoenvironmental data for the Australian interior arid zone are less abundant than for other parts of the country, proxy records from surrounding areas provide a reasonable Holocene chronology of moisture availability (Bowler 1976; Hesse et al. 2004; Ross et al. 1992). Generally speaking, aridity is enhanced during glacials and relaxed during interglacials. Lake level data from across the country show similar trends, with a low stand occurring at the LGM and water tables rebounding at ca. 10,000 $^{14}$C yr BP. Lakes reach a high stand around 6,500 to 5,500 BP $^{14}$C yr BP and decline. A minor high stand occurs again in some regions within the last 2,000 $^{14}$C yr BP (Hesse et al. 2004; Ross et al. 1992). In figure 3.11, I compare lithic ratios from Puntutjarpa to composite lake level diagrams from various regions of Australia. Though only a qualitative comparison, the similarity between lake level and artifact curves is striking, providing circumstantial support for the hypothesis that periods of aridity correlate with short stays at Puntutjarpa, while periods of relatively greater moisture availability correlate with occupations of longer duration. The curve formed by the local:nonlocal ratio shows considerably less noise than the debitage:nonlocal curve and suggests that the arid early Holocene was characterized by very brief stays, with the longest occupations occurring during the middle Holocene. The late Holocene was characterized by occupations of medium length. These findings provide strong support for the hypothesis that the local:nonlocal raw materials and debitage:nonlocal tools ratios primarily reflect mean per capita occupation span. Because Gould's twenty-three stratigraphic divisions do not represent twenty-three individual occupations, the patterning seen in the ratios through time appears independent of the number of occupations represented within each stratigraphic unit. Though changes in site

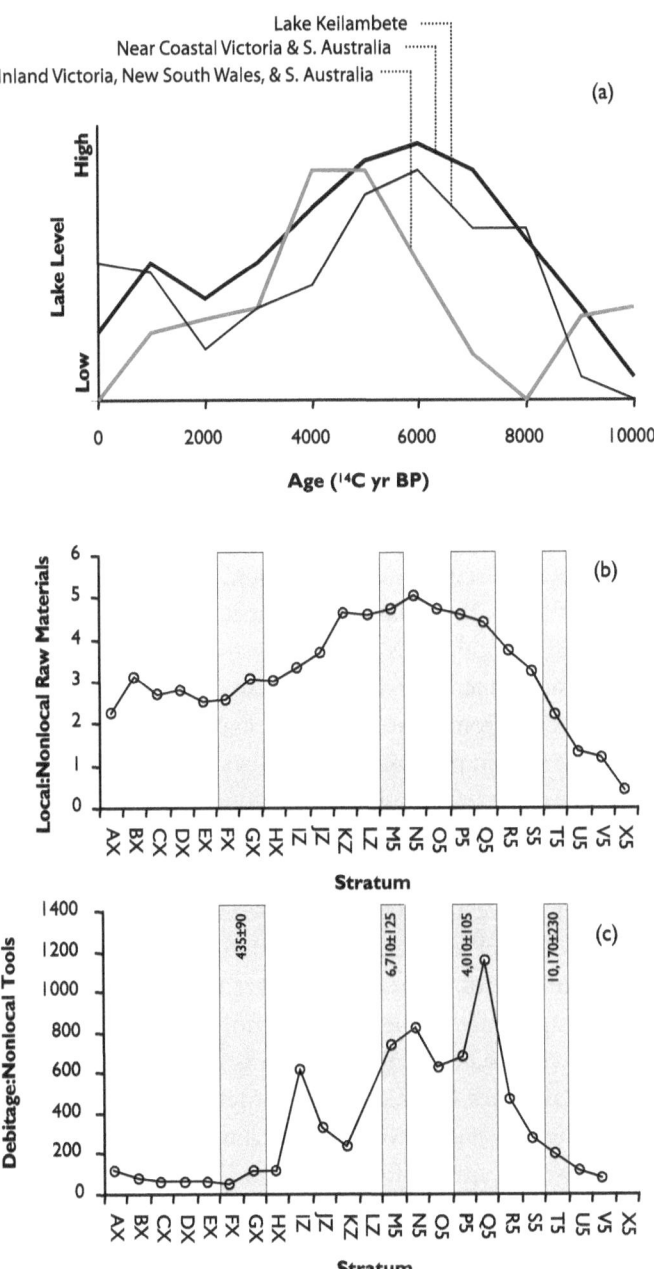

FIGURE 3.11. Comparison of (a) Holocene lake levels for various regions of Australia to (b) local:nonlocal raw materials, and (c) debitage:nonlocal tools from Puntutjarpa rockshelter. Radiocarbon dates from Puntutjarpa Trench 2 are shown as gray bars. Lake level figure redrafted from Ross et al. (1992). Puntutjarpa data are from Gould (1977).

function may account for subtle differences in the two ratios at different times, a strong mobility signal seems apparent.

## Conclusions

In this chapter, I developed methods for assessing the mean length of stay per site occupant. Specifically, I argue that the artifact type ratios can be used as measures of mean per capita occupation span if the discard rate of at least one artifact type changes as a function of occupation span. Because discard rates of artifacts transported into sites from previous residential moves are limited by the size of the transported toolkit, the ratios of local:nonlocal raw materials and debitage:nonlocal tools should provide proxy measures of the mean occupation span per site occupant. Furthermore, horizontal artifact densities should reflect the cumulative occupation span of all of the occupations present within an assemblage. Support for the first proposition is found in patterning in the data from the study sample supplemented with additional Paleoindian data from published sources. Data from Puntutjarpa shelter provide additional support. I examine the second proposition in the next chapter.

Notably, a range of mean per capita occupation spans is present in Folsom and Goshen sites. Although high-density Paleoindian sites are often "explained away" by arguments that they represent reoccupied quarry or workshop localities, the arguments developed in this chapter suggest instead that they do in fact represent relatively long-term occupations. It seems that Paleoindian settlement patterns were not as monotonous as they are sometimes portrayed, where groups moved rapidly from kill to kill, staying in a given location only briefly (e.g., Kelly and Todd 1988; Surovell 2000). Campsites like Carter/Kerr-McGee, Area 2 of Agate Basin, and Lake Theo do suggest very brief occupations, possibly associated with bison kills. In contrast, sites like Barger Gulch, Locality B, Bobtail Wolf, and possibly Hanson reflect a very different mobility strategy, one characterized by long occupation spans. Intermediate length occupations are evident as well in sites like Krmpotich and Mill Iron. How long is "long term"? This is an extremely difficult question to answer, but I will do my best to answer it in the next chapter.

Why would a group of hunter-gatherers, in particular large game specialists, remain in one place for an extended period of time? Kelly

(1995:111–160) uses a central-place foraging model and the marginal-value theorem (Charnov 1976b) to argue that hunter-gatherers should move camp when the potential foraging gain at a new location, minus the cost of moving to that location, outweighs the potential gains from staying in place. Two key variables to consider, then, are the relative "patchiness" of environments and the cost of moving. Kelly (1995:152) argues that sedentism "can be a product of local abundance in a context of regional scarcity." Using a different central-place foraging model, I reached similar conclusions based on the energetic costs of transporting food and children for mobile foragers (Surovell 2000). Kelly (1995:137–141) notes that the cost of moving is a function not only of the distance between suitable camping locations, but also terrain (e.g., frozen versus wet tundra, or dry ground versus deep snow) and the cost of dismantling and rebuilding housing. Similarly, the weight of loads to be carried during moves would also increase the costs of residential mobility.

These factors provide hints of why Paleoindians would occasionally break from stints of frequent residential mobility to remain stationary at particular locales. The increased patchiness of the environment and increased costs of mobility during winter make up one obvious seasonal explanation. Hunter-gatherers across the Great Plains and Rocky Mountains, and more generally hunter-gatherers in northern latitudes, commonly settle down in winter campsites (e.g., Binford 1991; Ewers 1955; Grinnell 1923; Smith 1974; Steward 1938). Presumably this broad phenomenon is a response to the increased costs of mobility in winter and/or increased environmental patchiness. Cold weather and snow accumulation bring greater mobility costs both in terms of energy and survival. Environments effectively become patchier in steppe and tundra biomes in the cold season because of increased requirements for fuel for fires. For northern hunter-gatherers, the distribution of suitable campsites becomes limited to areas where trees are present, or, in the case of some Inuit groups, locations where a reliable supply of sea mammal fat is available.

Therefore, I hypothesize that Folsom sites characterized by long-term mean occupation spans represent cold-season occupations (Greiser 1985). In contrast, Todd (1991) suggests that reduced seasonality during the late Pleistocene would have eliminated or lessened cold season mobility constraints. I argue that Paleoindians would not have necessarily permanently settled in single locations for the duration of the winter season, but that cold-season mobility was characterized by less frequent residential mobility because of the increased costs of mobility and greater spacing between suitable camping locations. In the specific case of Barger Gulch, located in the intermontane basin of Middle Park, an apt description of winter at the site 10,500 $^{14}$C years ago might have been "local abundance in a context of regional scarcity." It is a well-known phenomenon that ungulate densities in mountain valleys increase dramatically during the cold season as the animals are forced out of high elevations by deep snows. In Middle Park, mule deer and elk population densities increase dramatically during the winter (Carpenter et al. 1979), and prehistorically the same would have been true of bison. This is the "local abundance" part. The "regional scarcity" part comes from the relative isolation of the park during winter, where surrounding high mountains are characterized by many months of inhospitality. With the spring thaw and up-valley migration of ungulates, both the local abundance and regional scarcity would have been lessened, providing strong incentive to resume frequent residential relocation.

Unfortunately, very few independent data are available to test this hypothesis. There are little or no seasonality data for sites that appear to be characterized by lengthy occupations, usually because of poor bone preservation, as is the case for Barger Gulch and the Lake Ilo sites (Emerson 2000; Waguespack et al. 2002). Only one bison mandible fragment was recovered from Hanson, but it is suggestive of a late fall or winter occupation (Frison 1978:144). Also, based on a single calf mandible and the presence of fetal bone, the Area 2 Folsom component at Agate Basin appears to have been a late-winter or early-spring occupation (Hill 1994:39–40). While Agate Basin undoubtedly represents a relatively short-term occupation, the occupation span of Hanson remains unknown. Nonetheless, the evidence from Agate Basin may suggest that Paleoindians did, at least occasionally, engage in frequent residential mobility during winter or early spring.

# 4

# The Reoccupation Problem

ARCHAEOLOGISTS HAVE long recognized that single archaeological components could be composed of multiple occupations. For example, when Emil Haury (1959:27) was faced with a Clovis site containing remains of at least nine individual mammoths, he asked, "Were they all killed at once?" In this case, the possibility of reoccupation in the Clovis level at the Lehner site was clear. To Haury, it seemed only a remote possibility that Clovis hunter-gatherers could have succeeded in killing nine mammoths in one hunt and in having those mammoths fall in close proximity to one another (but see Saunders 1977). Rarely, though, are we faced with such scenarios where the archaeological record provides us with a clear suggestion of reoccupation when no stratigraphic separation of occupations is evident.

Reoccupation of archaeological sites is a pattern established in the earliest of all archaeological sites (Semaw 2000), so reoccupation of sites must always be considered a possibility. For Haury, it was the assumption that all of the mammoths at Lehner fell prey to Clovis hunters that led him to question whether a single event was represented. For Ingbar (1992, 1994), I would argue, it was the assumption that Northern Plains Folsom sites should look like Southern Plains Folsom sites that eventually led him to the conclusion that Hanson was the product of "many occupation events" (Ingbar 1992:186). Of course, all models, whether formal or informal, are built from a set of assumptions. In this chapter, working from a set of assumptions and building on the measures of mean per capita occupation span developed in chapter 3, I construct a simple model of lithic accumulation that shows how one might distinguish single occupations from reoccupied sites. This model follows from those of Lightfoot and Jewett (1984, 1986) and Gallivan (2002), which were discussed in the previous chapter.

## Detecting Reoccupation in Archaeological Sites

For any set of positive numbers, its mean is less than its sum (go ahead, try it). For a single number, its "mean" is equal to its sum. It is this simple

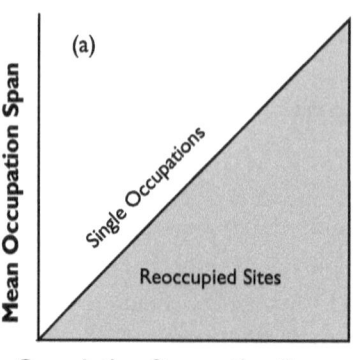 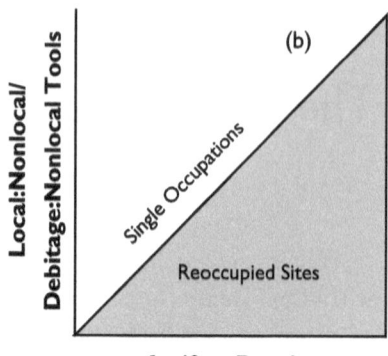

FIGURE 4.1. (a) A model for distinguishing between single occupations and multiple occupations using the relationship between cumulative occupation span (x-axis) and mean occupation span (y-axis). (b) Same with proxy variables substituted for generalized variables.

mathematical fact that can be used as a basis for detecting the reoccupation of archaeological sites, because for reoccupied sites (n occupations $>1$) the mean occupation span will be less than the cumulative occupation span. Consider three occupations lasting 3, 9, and 15 days. The cumulative (summed) occupation span is 27 days. The mean occupation span is 9 days. Contrast this with a single occupation of 27 days. The mean and sum are 27. If one were to plot cumulative versus mean occupation span for a set of single occupation sites (fig. 4.1a), the results would fall on a line with a slope of one and a y-intercept of zero ($y = x$). Reoccupied sites would fall to the right of that line ($y < x$).

As argued in chapter 3, ratios of local:nonlocal raw materials and debitage:nonlocal tools can provide proxy measures of mean per capita occupation span. Because measures of mean per capita occupation span monitor the average length of stay for individuals, deviations from the line should indicate reoccupation by individuals or groups. Fundamentally, such events are identical but operate at slightly different scales. Unless group membership is extremely variable, such deviations, I would argue, are likely a product of multiple occupations by groups rather than individuals since the impact on an assemblage by the coming or going of one or two individuals should be slight compared to that of multiple occupation by large numbers of people. Horizontal artifact densities can be used as a proxy for cumulative occupation span since the occurrence of multiple occupations leads to greater numbers of artifacts per unit area (fig. 4.1b). Generally speaking, reoccupied sites should be characterized by high artifact densities coupled with relatively low frequencies of local raw materials and debitage.

Applying the model archaeologically is not simple because mean and cumulative occupation spans are not known and can be estimated only using proxy variables, meaning that the exact expected relationship between the variables for a set of single occupations is unknown. Therefore, a correlation between cumulative and mean occupation span could indicate a set of single occupations or a set of sites that show similar degrees of reoccupation. Furthermore, these ratios might be expected to vary considerably because of differences in sampling (e.g., screen mesh size), reduction (e.g., bifacial vs. blocky core), the state of the toolkit on arrival, and other factors. Nonetheless, this model is at least a point of departure for attempting to solve this difficult problem. No doubt it will be refined with larger samples and additional lines of evidence. One way to test the model, however, is to apply it to a site known to have experienced multiple occupations. Combining multiple levels from a single site should cause the point scatter formed by the two proxy variables to migrate to the right. Thus, I begin with data from Puntutjarpa shelter (Gould 1977) and later turn to the Paleoindian data set from the study sample.

## Testing the Model: Puntutjarpa Rockshelter

To calculate artifact densities by level for Puntutjarpa rockshelter, it was first necessary to estimate excavation areas by level. At Puntutjarpa, the irregular back wall of the shelter formed the southern boundary of the excavation (Gould 1977:fig. 33). Because Gould mapped the horizontal position of the back wall at one-foot depth intervals, I was able to estimate excavation area for most levels with a reasonable degree of accuracy. Excavation areas for levels falling between the one-foot contour intervals were estimated by interpolation. Table 3.4 shows the excavation area and artifact density estimates by stratigraphic level for the site.

In figure 4.2, I compare artifact density to three measures of mean per capita occupation span: the local:nonlocal and debitage:nonlocal tool ratios, and the occupation span index (OSI) (table 3.4). The OSI averages

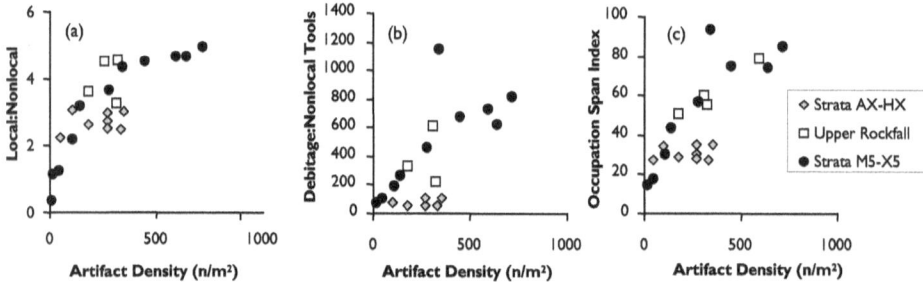

FIGURE 4.2. Artifact density vs. (a) local:nonlocal raw materials, (b) debitage:nonlocal tools, (c) occupation span index for Puntutjarpa rockshelter. Markers are shown by stratigraphic level.

adjusted values of the two ratios to combine them into a single measure to reduce noise. To calculate the OSI, values for each ratio are transformed to values ranging from 0 to 100 by standardizing each observation to the largest for that variable and multiplying by 100. The two values are then averaged. This transformation is performed because values for the two ratios are scaled very differently at Puntutjarpa (table 3.4), and using an average of the untransformed values would heavily skew the OSI toward the debitage:nonlocal tool ratio. What is immediately apparent from figure 4.2 is that all measures of mean per capita occupation span are correlated with artifact density, providing additional support for the hypothesis that the two ratios do reflect occupation span. Importantly, artifact density and occupation span measures are largely independent, statistically speaking.

Because the expected relationship for a group of single occupations is unknown, it is not possible to determine which if any stratigraphic levels represent single occupations, but the degree of reoccupation represented by each level should be reflected by its position on the x-axis. It is also important to note that the relationship to artifact density for each measure appears to take a slightly different form. Linear, log-linear, exponential, and power functions can be fit to each distribution. The greatest degree of correlation between artifact density and OSI occurs with a power function (OSI = 4.26 • density$^{0.4272}$, $r^2$ = .6354, $p < 0.001$). This may indicate problems with some of the simplifying assumptions made in the previous chapter, a point I return to later.

To demonstrate that the framework for detecting reoccupation shown in figure 4.1 is valid, it is a simple matter to combine data from multiple

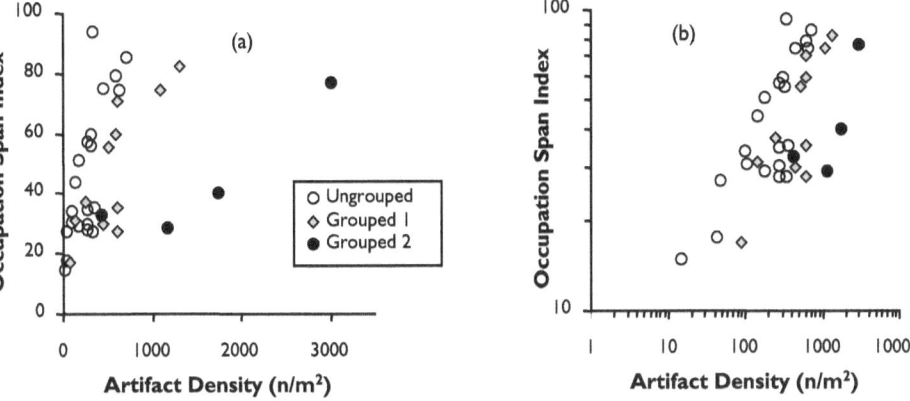

FIGURE 4.3. (a) Artifact density vs. occupation span index for ungrouped and grouped levels from Puntutjarpa rockshelter. (b) Same with both axes logged. Ungrouped data are for the twenty-one stratigraphic levels for which the OSI can be calculated. Grouped 1 are the data lumped into groups of two or three stratigraphic levels. Grouped 2 are the data lumped into groups of five or six stratigraphic levels.

stratigraphic levels to "simulate" reoccupation. When this is done, the point scatter should migrate to the right. Excavation areas for each set of grouped levels were averaged in the density calculation. In the first iteration, the twenty-three stratigraphic levels were combined into adjacent groups of two, but because of an odd number of levels, the last three levels (U5–X5) were combined into a single grouping. In the second iteration, the levels were grouped into adjacent groups of six and one group of five (S5–X5). As predicted, with each successive grouping the point scatter moves to the right (fig. 4.3). This is not a surprising result since with each grouping densities become summed and artifact ratios are averaged. In figure 4.3b, both axes are logged, showing the emergent and expected wedge-shaped distribution formed by assemblages known to represent reoccupation.

## Applying the Model: The Study Sample

Having established the validity of the model, it is now possible to apply the same test to the assemblages of the study sample. Again, the predicted relationship between artifact density and the OSI is unknown. Therefore, without a very large sample, it is unlikely that it will be possible to determine

TABLE 4.1. Occupation span index and horizontal artifact densities for the sites of the study sample for all artifacts larger than 2 cm.

| Variable | Site | | | | |
|---|---|---|---|---|---|
| | CKM | AGB | KRMP | BGB | UTM |
| Local:nonlocal | 0.09 | 0.69 | 14.38 | 240.14 | 5 |
| Debitage:nonlocal tools | 1.47 | 5.96 | 31.81 | 302.8 | 2.50 |
| Occupation span index (OSI) | 0.26 | 1.13 | 8.25 | 100 | 1.45 |
| Ln (OSI) | −1.35 | 0.12 | 2.11 | 4.61 | 0.37 |
| Total artifacts (>2 cm) | 40 | 532 | 584 | 1,688 | 12 |
| Excavation area (m$^2$) | 34† | 171‡ | 74 | 51 | 25.5 |
| Artifact density (artifacts/m$^2$) | 1.18 | 3.11 | 7.89 | 33.10 | 0.47 |
| Ln (artifact density) | 0.17 | 1.13 | 2.07 | 3.50 | −0.76 |

*Notes:* †Excavation area for Carter/Kerr-McGee is based on unpublished data. Although the published map of the excavation block shows an area of approximately 285 m$^2$, the artifacts that could be confidently assigned to the Folsom component are found in two isolated portions of the excavation. Therefore, a reduced estimate of 34 m$^2$ based on actual artifact locations was used. ‡The excavation area for Agate Basin, Area 2 was estimated from Frison's (1982b) plan map and differs significantly from Hill's (1994:9) estimate of 121 m$^2$.

if the sites of the study sample represent single occupations. If all are single occupations, however, there should be a strong correlation between artifact density and OSI. If no correlation is found, it is likely that at least one assemblage represents the cumulative effects of multiple occupations.

To control for potential sample biases, only artifacts larger than 2 cm were used in this analysis. Artifact density and OSI for each site are shown in table 4.1 and figure 4.4. As at Puntutjarpa, artifact density and OSI are highly correlated for the five sites of the study sample (linear correlation, $r^2 = 0.979$; $p = .001$). However, this correlation is largely driven by Barger Gulch, an outlier in comparison to the remaining four sites (fig. 4.4a). After logarithmic transformation (base e) of artifact density and OSI (fig. 4.4b), the outlier problem is eliminated, but the correlation is not statistically significant ($r^2 = 0.738$; $p = 0.062$). However, Upper Twin Mountain again breaks from the trend defined by the campsites, and if it is excluded, there is a very strong correlation between the log-transformed values ($r^2 = 0.997$; $p = 001$). Therefore, virtually all of the variance in the campsite OSI can be explained by artifact density, and there is approximately a one in one thousand probability that this relationship can be explained by

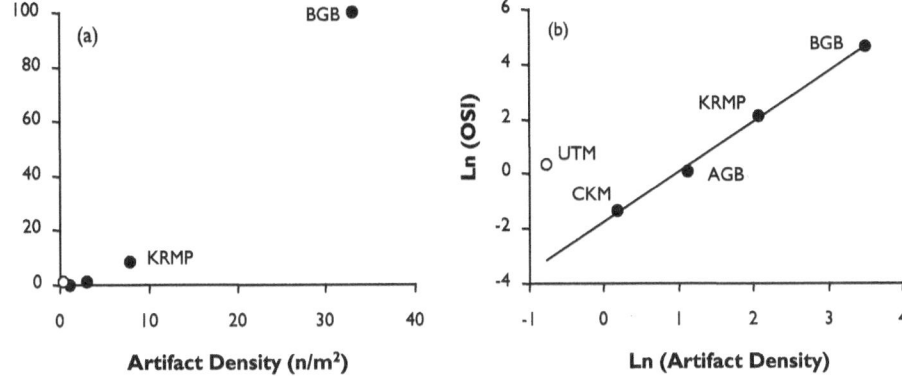

FIGURE 4.4. (a) Artifact density vs. occupation span index for the study sample. White marker is Upper Twin Mountain (kill site). Black markers are campsites. (b) Same for log transformed (base e) data. Regression line for campsite, excluding Upper Twin Mountain (y = 1.8174x − 1.7489; r² = 0.997; p = 0.001).

chance. This provides support for the hypothesis that the assemblages from the four campsites of the study sample represent single occupations. However, there remains a possibility, though somewhat unlikely in my mind, that all four campsites are reoccupied to the same degree. To distinguish between these possibilities would require a larger sample.

Although a linear trend line could be fit to the log-transformed campsite data, the relationship between artifact density and the OSI for the study sample, like Puntutjarpa, is best described by a power function: OSI = 0.1753 (Density)$^{1.8142}$. It can be shown, however, that the accumulation models developed in chapter 3 predict a near linear trend between artifact density and OSI. There are many ways to account for this discrepancy. For example, it is possible that the four campsites do not represent single occupations and that fitting a power function to this scatter is inappropriate. Given the strength of the observed correlation, however, this seems unlikely. Mathematically, the simplest solution to the problem is to modify the function relating artifact density to occupation span. If equation 3.7 relating artifact density ($d/s_o$) to occupation span ($t$) is instead modified to increase as a power function of occupation span where the exponent is less than one:

$$\frac{d}{s_o} = \frac{rt^x}{a}, \quad x < 1 \qquad (4.1)$$

the observed pattern can be explained by the model. This modification of the artifact density model is supported by Yellen's (1977:129–130) observation that the absolute size of the scatter left at !Kung campsites increases with greater occupation span. Presumably, as occupations are lengthened, the scale of scatters increases because of cleaning and other behaviors causing artifact dispersal, thereby causing the rate of increase of artifact density to decline with time.

## Estimating Actual Occupation Span

If all four campsites represent single occupations, it should be possible to produce a semiquantitative estimate of actual occupation span. Without knowing discard rates, however, any derived estimates must be viewed with a fair degree of uncertainty. What can be said with a high degree of certainty is that the Carter/Kerr-McGee Folsom occupation is characterized by the shortest occupation span (both mean and cumulative) followed by Agate Basin Area 2, Krmpotich, and Barger Gulch, Locality B, respectively. It should be noted that if discard rates differed dramatically among these sites, even this ordinal ranking could be incorrect. Lacking a measure of occupation span that is independent of discard rate, however, this is a hypothesis that is at the moment very difficult to test.

Both artifact densities and the OSI could be used to estimate occupation span, but here I rely solely on artifact density. Although for a large set of sites OSI should increase almost linearly with occupation span, this will not be true for sites characterized by brief occupations where the function takes a subtle nonlinear form (fig. 3.4a). Likewise, as argued above, artifact densities may not increase linearly with occupation span, making these estimates even more suspect. To develop a semiquantitative estimate of actual occupation, I standardized artifact densities to Carter/Kerr-McGee, the site characterized by the shortest occupation span. Artifact density at Barger Gulch, for example, is 28.1 times greater than that of Carter/Kerr-McGee. If a particular length of occupation is assumed for Carter/Kerr-McGee, the occupation span for the remaining sites can be estimated by multiplying by the standardized density. For example, if a one-day occupation is assumed for CKM, the Folsom occupation at Area 2 of Agate Basin is estimated to have lasted 2.6 days, the Krmpotich occupation lasted approximately one week, and the Barger

Gulch occupation lasted twenty-eight days, or almost one month. If a one-week occupation is assumed for CKM, the estimate for Barger increases to 196 days, or more than six months. Although it is possible that these estimates are correct, differences in discard rates among the sites are likely skewing them. I explore this issue further in chapter 8, but for the moment the important point is that some Folsom sites appear to be characterized by fairly long-term occupations. Even minimal estimates for the Folsom occupation at CKM lead to estimates of approximately one month for the occupation of Barger Gulch.

*Discussion*

It may seem unlikely to some that all four campsites of the study sample represent single occupations. After all, three of these sites are known to have been reoccupied at some time during the Paleoindian period. At Carter/Kerr-McGee, overlying the Folsom component are Hell Gap/Agate Basin and Cody components. It is also underlain by a Clovis or Goshen component (Frison 1984). Likewise, the Agate Basin site contains many loci of Paleoindian occupation and some of those (including Area 2) exhibit multiple Paleoindian occupations. Area 3 even exhibits two vertically differentiable Folsom components (Frison 1982b). At Barger Gulch, separated horizontally in space, are at least ten Paleoindian occupations (Kornfeld et al. 2001; Waguespack et al. 2002, 2006). At Locality B alone two distinct surface clusters of Folsom artifacts are present, possibly indicating two separate occupations (Waguespack et al. 2002). This pattern is not unique to the study sample but is repeated at numerous Paleoindian sites (e.g., Bement 1999; Forbis and Sperry 1952; Frison and Bradley 1980; Hester 1972; Irwin-Williams et al. 1973; Johnson 1987; Wilmsen and Roberts 1984).

What this finding seems to suggest is that although Paleoindian foragers regularly reused certain locations on the landscape, only after an extended time had passed would they reoccupy the exact same location. A couple of cautionary notes are warranted here. First, very brief reoccupations by few individuals would likely go undetected using these methods, particularly for sites characterized by relatively long-term mean per capita occupation spans, because the later contribution to the assemblage would be swamped by the earlier more intensive occupation.

Second, I do not mean to imply that all Folsom components are "pure," in the sense of being characterized by only a single occupation. For example, if Hanson or Agate Basin, Area 3, each containing multiple Folsom components, were deflated onto a single surface, the two Folsom components would be merged, and such events must have occurred. Nonetheless, if I am correct in concluding that only single occupations are present at Locality B of Barger Gulch, Krmpotich, Area 2 of Agate Basin, and Carter/Kerr-McGee, it is interesting to speculate as to why.

One explanation is that the use of a specific campsite reduces its potential for future habitation because of resource depletion in the immediate area and/or an accumulation of debris and waste (Binford 1983). The depletion of firewood, in particular, could make a location unsuitable for future habitation for many years because it would have a relatively long regeneration time (Moore 1987:143–144). Obviously, the impact of resource depletion and trash accumulation on campsite suitability will be proportional to occupation intensity (people × time). A second and perhaps more likely explanation is that spatially congruent reoccupation is a very low probability event because there could be many suitable camping locations, even in very attractive areas of the landscape where reliable resources are available (e.g., perennial streams in arid regions). For example, if 50 hectares are optimally suited for residential occupation at a site like Barger Gulch and an average Folsom camp requires 0.25 hectares of space, how many randomly located Folsom occupations are necessary before two of those are likely to overlap in space?

To answer this question, I wrote a simple simulation. In this simulation, circular campsites are randomly placed within a square area. The campsites can be located anywhere within the area, but they are not permitted to intersect with its edge. Each simulation was run for 10,000 iterations to determine the probability of at least two campsites overlapping and the number of overlaps that occur (fig. 4.5). For 50 ha of available space and circular campsites of 0.25 ha in area, the probability of having at least one overlap exceeds 50 percent at seventeen occupations.

Using this estimate, if the Folsom period lasted 800 years and such a location is reoccupied once every 50 years, then approximately sixteen campsites would be represented, and, on average, zero or two would likely overlap. In other words, even at large reoccupied sites, such as Barger Gulch, Blackwater Draw, and Hell Gap, the probability of overlapping

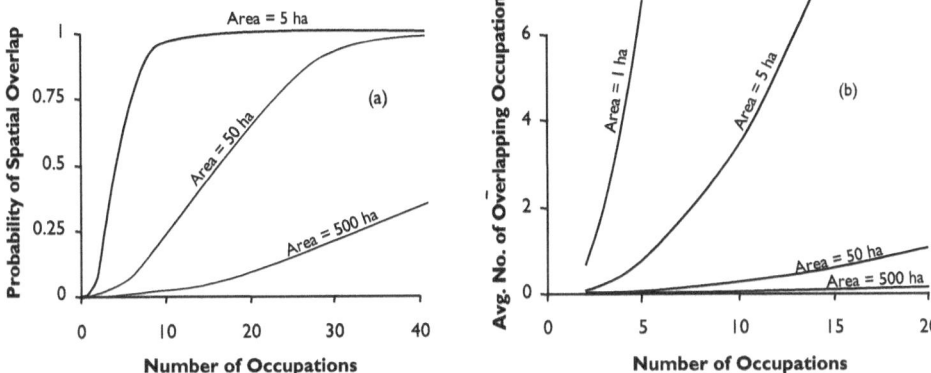

FIGURE 4.5. (a) The number of occupations vs. the probability of at least two spatially overlapping occupations for camping areas of different sizes. (b) The number of occupations vs. the average number of spatially overlapping occupations for camping areas of different sizes. All probabilities were estimated by simulation.

occupations within the time frame of a single Paleoindian complex might be relatively small. The most critical variables are the size of the available camping area, the size of sites, and the frequency of occupation. Cave sites, like Puntutjarpa, should and do show repeated horizontally overlapping occupation because the area available for habitation is limited. In large open-air sites, where plenty of room might be available for residence, it may take many hundreds of years before two occupations overlap in space. Furthermore, two occupations that overlap only slightly should not register a signal of reoccupation as strongly as two that overlap considerably. With these considerations, perhaps the finding that these four Folsom components each represents single occupations should be no surprise at all.

# 5

# Stone Age Supply-Side Economics

IN 1992 WHEN BILL CLINTON was elected to the office of president of the United States, the difference between federal budget revenues and expenditures equated to a $340 billion deficit (fig. 5.1). At that time, the public held just over $3 trillion in debt. During the Clinton Administration, the budget deficit gradually shrank. After decades of the government spending more than it collected, in 1999 revenues exceeded spending, and the country enjoyed a $1.8 billion budget surplus. By the year 2000, the budget surplus increased to $86.6 billion, coincident with the 2000 presidential election. The ensuing political debate, strangely enough, centered on what to do with our newfound budget prosperity.

In campaign rallies, Bush promoted his doctrine that most of the surplus should be given back to the people who created it, the American taxpayers. More money for Americans meant greater economic growth. While Gore also advocated tax cuts, he argued for a lower but targeted tax cut to low- and middle-income families who stood to benefit most. Both candidates also introduced new spending proposals, though they disagreed on how the extra funds would be best allocated. Likewise, both candidates advocated paying down the national debt. Gore alone proposed the creation of a $300 billion "rainy day fund" to safeguard against the possibility that the surplus projected by the Congressional Budget Office would not materialize. Only a few years later, we found ourselves once again producing debt at record rates, with budget deficits exceeding $500 billion (fig. 5.1).

Budget economics are not new to North America. Prehistoric humans faced similar issues day in and day out, although the currencies involved were quite different. Deficits of food could be deadly, but accumulating surpluses of food could be costly. The same could be said of tool stone. Tool stone is a resource that was regularly used by prehistoric hunter-gatherers. The amount of tool stone on hand at a residential location is finite. Therefore, to maintain a regular supply of stone to fulfill demand,

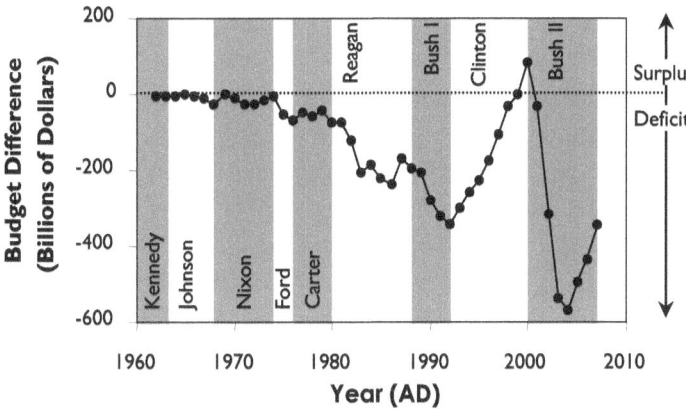

FIGURE 5.1. The U.S. federal budget deficit or surplus for the fiscal years 1962 to 2007. Data from the U.S. Congressional Budget Office.

prehistoric humans would necessarily travel to raw material source areas and return with lithic raw materials to be allocated to tools for the performance of various tasks. Should the rate of supply exceed the rate of demand to create surpluses of raw materials? Should people bring back more stone than is necessary for their immediate needs? In sites we commonly find artifacts that have been abandoned well before they have reached the end of their potential use-lives. Residential sites that contain large cores, for example, imply the apparently wasteful transport of lithic raw materials from a source area. I say "wasteful" because it requires time and energy to accumulate a surplus, but no apparent gains are made on that labor investment if the extra raw material is simply abandoned. This implies that there must be some benefit to the accumulation of lithic surpluses.

Returning to the 2000 political debate, we can explore the costs and benefits of maintaining surpluses in budgets, whether lithic or monetary. Bush and Gore both advocated tax cuts because there are potential economic costs to the nation if the government extracts more revenue from its citizens than is necessary for the day-to-day operation of the government. This observation would suggest the need to eliminate or at least to place an upper limit on surpluses, so the economy is not bled into recession. There are, however, benefits to surpluses. Both Bush and Gore promoted spending programs that would invest money in the country, which in turn should provide higher returns and higher revenues. Gore alone suggested another benefit. Because there is some uncertainty in supply (i.e., the Congressional Budget Office can make only rough estimates of expected tax revenues), maintaining a budget surplus could cushion us against a possible deficit if revenues fall short or spending exceeds expectations. The possibility of natural disasters, fires, nuclear meltdowns, and terrorist attacks coupled with varying participation in government programs means the exact amount the government will spend can only be estimated. Therefore, maintaining a surplus provides a cushion against the possibility the government will outspend its revenues. Using such costs and benefits, it is possible to develop a formal model of surplus accumulation.

This chapter is not about politics. Instead, it is about economics, as issues of budgets and surpluses have wide applications to a broad array of currencies, such as money, food, water, firewood, or raw materials for tool manufacture. This is not to say that a single model can be applied to any of these currencies without modification. The breadth of its application is limited by the nature of its assumptions, but it can serve as useful foundation on which to build more complex or alternative models. In this chapter, I develop and test a model of lithic economy that is intended to answer the questions: Why is it beneficial to maintain a budget surplus, and how much of a surplus should be maintained?

## The "Rainy Day" Model

Time and energy are expended in provisioning residential locations with lithic raw materials, and optimizing foragers should behave so as to maximize the return on this investment of labor. In this respect, the transport of a surplus of lithic raw materials to a place seems wasteful because no return is gained from the extra effort made to accumulate that stone. Therefore, we might expect humans to acquire raw materials only when they are needed and therefore eliminate the possibility of investing more labor into lithic supply than is required by the level of demand. If, however, the rate of consumption and/or supply of lithic raw materials are somewhat unpredictable, then it may be optimal to maintain a surplus because unexpected shortfalls will require extra trips to procure lithic raw material, and the costs of being caught short may be very high. Like Gore's "rainy day fund," the maintenance of a surplus can cushion stone tool users against the possibility of lithic budget deficits.

The following model of lithic surplus accumulation is based on this idea, and it concerns the optimal surplus size for an individual provisioning a campsite with lithic raw materials. Assume the following:

- $u$ = the mean per capita rate of raw material consumption (e.g., kg/person/day)
- $a$ = the mean per capita rate of stone procurement (e.g., kg/person/day)
- $s$ = the size of the surplus to be created (e.g., in kg)
- $t$ = mean per capita occupation span (e.g., in days)

The size of the lithic surplus as a function of time ($t$) is calculated as the difference between the amount of stone procured and consumed:

$$s = at - ut \tag{5.1}$$

The rate of procurement ($a$), then, can be expressed as a function of the size of the surplus ($s$), the rate of consumption ($u$), and the occupation span ($t$):

$$a = \frac{s}{t} - u \tag{5.2}$$

The utility of this expression becomes apparent once a few more assumptions are made:

- $e$ = the average cost (e.g., kcal/kg) of acquiring lithic raw materials in embedded procurement. I assume that stone is normally acquired, when provisioning a location and accumulating a surplus, via embedded procurement, because in most cases it is less costly than direct procurement (Binford 1979).
- $d$ = the average cost (e.g., kcal/kg) of acquiring lithic raw materials in direct procurement. I assume that shortfalls prompt immediate raw material procurement in the form of direct forays to raw material source areas. The costs of direct and embedded procurement could be expressed in terms of energy or time and are intended to include both resource and opportunity costs (Hames 1992).
- $p$ = the probability of suffering a shortfall in lithic raw materials per unit time.
- $x$ = the average amount of stone transported following a shortfall (e.g., kg)

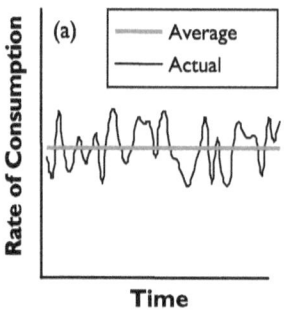 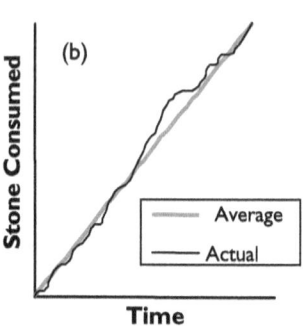

FIGURE 5.2. (a) A model of actual and average rate of consumption of lithic raw materials for an individual over the span of an occupation. Because the quality and quantity of tasks to be performed will vary from day to day, the rate at which stone is used should fluctuate around some mean. (b) The total quantity of stone used as a function of time assuming the average and the actual rates of consumption shown in (a).

To maintain a constant supply of lithic raw materials, the rate of stone procurement must minimally keep up with the rate of consumption. Therefore, the work ($w$) expended in transporting stone to a campsite of occupation length $t$ when supply perfectly matches demand is calculated as:

$$w = eut \tag{5.3}$$

Because I use the mean rate of consumption of stone, the average amount of stone used as function of time will be a linear function of time. However, we can expect the actual amount of stone used to vary around that mean since there should be some variability in the amount of stone used daily owing to changes in the amount and type of work performed (fig. 5.2). For example, in gearing up for a bison hunt a group of foragers may find that they require more lithic raw material than is currently on hand. Therefore, an extra trip to procure additional raw material might be necessary. In that case, a forager may find it advantageous to acquire stone at a rate that exceeds the average rate of use, thereby creating lithic surplus. Therefore, equation 5.3 is rewritten with respect to the rate of stone acquisition:

$$w = eat \tag{5.4}$$

However, we are not interested in the rate of stone procurement as much as the size of the surplus created. By plugging equation 5.2 into equation 5.4, work is expressed as a function of surplus size:

$$w = et\left(\frac{s}{t} + u\right)$$
$$w = se + uet \qquad (5.5)$$

In this formulation, there would be no advantage to maintaining a surplus at all because work would be minimized if $s = 0$. For the creation of a surplus to be advantageous, a final term representing the cost of lithic deficits or shortfalls must be introduced:

$$w = se + uet + ptxd \qquad (5.6)$$

where $pt$ represents the number of shortfalls that will occur for an occupation span of length $t$, $x$ is the average amount of stone acquired during a shortfall, and $d$ represents the mean cost of procuring raw material per unit mass in response to those shortfalls.

The maintenance of a surplus will cushion against the possibility of lithic budget shortfalls because the probability of a shortfall ($p$) should be inversely related to the size of the surplus. If, for example, no surplus is maintained ($s = 0$), the probability of shortfalls will be high, but if a large surplus is on hand, the probability of a shortfall will be low. Therefore, $p$ is expressed as a function of $s$:

$$p = \frac{k}{s + k} \qquad (5.7)$$

where $k$ is a proportionality constant representing the scale of variation in rates of raw material acquisition (supply) and consumption (demand). This function was used because it limits the range of $p$ to values between zero and one.[1] Figure 5.3 graphically depicts equation 5.7 with various values of $k$. At small values of $k$, the probability of a shortfall drops rapidly with even a small surplus. At larger values of $k$, the probability of a shortfall declines more slowly, and shortfalls remain likely even with relatively large surpluses. A simple way to conceptualize the variable $k$ is that it represents the degree of stochasticity or predictability in rates of consumption. When the amount of stone to be used

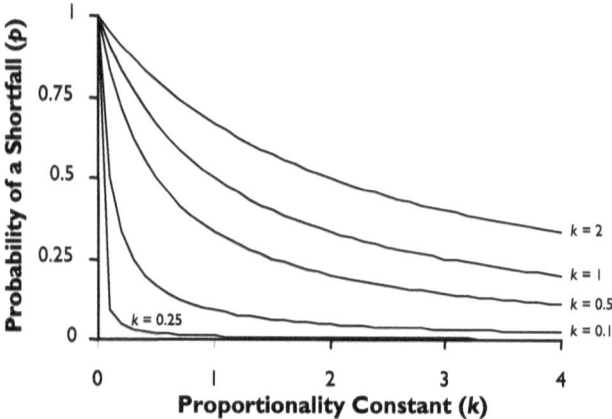

FIGURE 5.3. Graph of equation 5.7 showing the probability of a shortfall ($p$) as a function of the size of the surplus ($s$) for various values of the proportionality constant ($k$).

per day is extremely variable and unpredictable, larger surpluses will be necessary to prevent shortfalls. Substituting equation 5.7 for $p$ in equation 5.6 yields:

$$w = se + uet + \frac{k}{s+k} xtd \qquad (5.8)$$

To allow comparison of surplus sizes across a set of sites or components of varying occupation length, equation 5.8 must be divided by occupation span ($t$) to provide the final currency of cost per unit time:

$$\frac{w}{t} = \frac{se + uet + \frac{k}{s+k} xtd}{t}$$

$$\frac{w}{t} = \frac{se}{t} + ue + \frac{kxd}{s+k} \qquad (5.9)$$

In equation 5.9, a trade-off is apparent. In the first term, a larger surplus equates to more energy expended per unit time in accumulating a surplus because $s$ is in the numerator. In the third term, because $s$ is in the denominator, larger surpluses reduce the work spent per unit time in the acquisition of raw material following shortfalls. A surplus that is too small will not effectively buffer you against the possibility of lithic shortfalls, but too large a surplus may be very costly to accumulate. Therefore, we

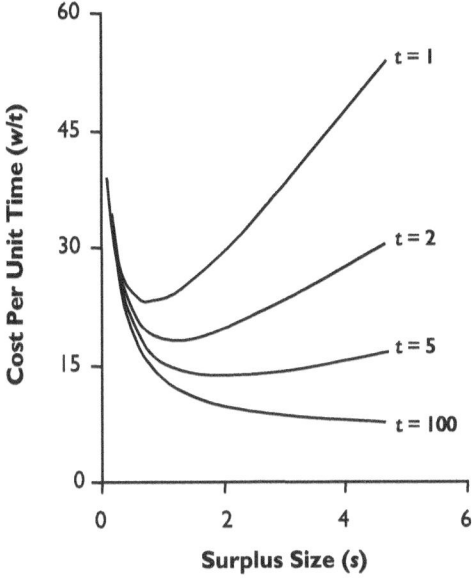

FIGURE 5.4. Graph of equation 5.9 showing optimum surplus sizes for various occupation spans ($t$), assuming $k = 0.2$, $u = 0.5$, $d = 50$, $x = 1$, and $e = 10$.

can expect there to be some optimum surplus size given the constraints in place (fig. 5.4).

Figure 5.4 shows two primary effects of occupation length ($t$) on the curve defining optimum surplus size: (1) Larger surpluses are expected in longer occupations; (2) while it is very costly to accumulate large surpluses in short-term occupations, it is relatively inexpensive to do so in long-term occupations. Viewing the latter effect from the first term of equation 5.9 ($se/t$), note that the cost of surplus accumulation decreases as occupations are lengthened. For example, if it costs 100 kcal to procure 1 kg of lithic raw material, the cost to accumulate a 10-kg surplus of raw material in a one-day occupation would be 1,000 kcal/day. If the same surplus were accumulated over the duration of a 100-day occupation, the cost would be only 10 kcal/day. We can frame the same problem in a different way. Assume a forager always creates a 10 kg surplus of raw materials at a site regardless of occupation span. If that forager moves daily over the course of a year, 3,650 kg of raw material will have been procured but not used. If that forager moves only once annually, only 10 kg of stone will go to waste.

To determine which variables govern optimal surplus size, the first derivative of equation 5.9 is first solved with respect to $s$:

$$\frac{d(w/t)}{ds} = \frac{e}{t} - \frac{dkx}{(k+s)^2} \qquad (5.10)$$

The optimum surplus size ($s_{opt}$) is calculated by solving equation 5.7 for $s$ when $d(w/t)/ds = 0$:

$$0 = \frac{e}{t} - \frac{dkx}{(k+s_{opt})^2}$$

$$s_{opt} = \sqrt{\frac{d}{e} xkt} - k \qquad (5.11)$$

From equation 5.11, it is apparent that the decision of how large a surplus should be based on only five variables: the occupation span ($t$), the relative cost of direct to embedded procurement ($d/e$), the average amount of stone acquired during a lithic shortfall ($x$), and the proportionality constant ($k$). If $k$, $d$, $x$, and $e$ are constant, again, we would expect larger lithic surpluses to be present in sites with longer occupation spans, and the size of the surplus should increase as a function of the square root of occupation span (fig. 5.5). The same relationship will hold for the relative costs of direct and embedded procurement ($d:e$) and the amount of stone acquired during shortfalls ($x$). As the cost of direct procurement increases, or the cost of embedded procurement decreases, the optimum surplus size will increase as a square root function of their ratio. Another prediction of the model is that as $k$ increases, optimal surplus size will increase to a point and then start to decline (fig. 5.6). Essentially, if $k$ gets so large (rates of consumption are very unpredictable) that maintaining a surplus does little to insure you against lithic shortfalls, the optimal surplus size will decline. That is, if a large surplus does not prevent deficits, then by creating a surplus you will incur both the costs of surplus accumulation and deficit costs. In that case, you are better off by acquiring raw material in direct response to shortfalls. In practice, however, it is difficult to conceive of a situation in which rates of raw material use would be so stochastic that the maintenance of some surplus would not provide a benefit.

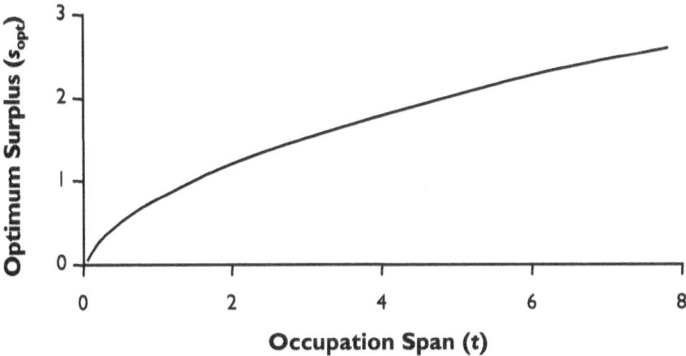

FIGURE 5.5. Graph of equation 5.11 showing the relationship between occupation span ($t$) and the optimal surplus size ($s_{opt}$), assuming $k = 0.2$, $d = 50$, $e = 10$, and $x = 1$.

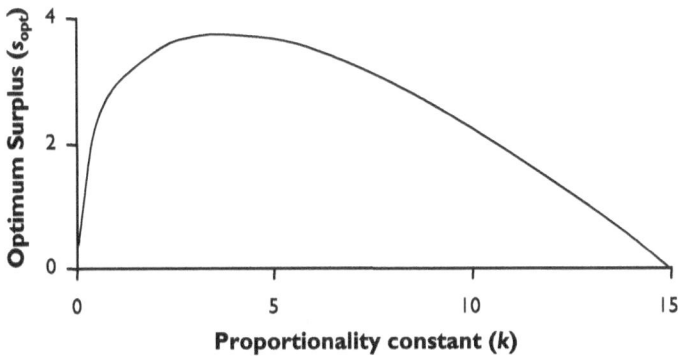

FIGURE 5.6. Graph of equation 5.11 showing the relationship between the proportionality constant ($k$) and optimal surplus size ($s_{opt}$), assuming $t = 3$, $d = 50$, $e = 10$, and $x = 1$.

## Testing the Model: Occupation Span versus Surplus Size

The model predicts that larger surpluses are expected with longer occupation spans. It also predicts that the relative costs of direct and embedded procurement will impact surplus size. Because the time and energy expended in direct and embedded procurement should be a function of the distribution of lithic raw material sources relative to residential locations, it is difficult to test the model using a series of sites where the

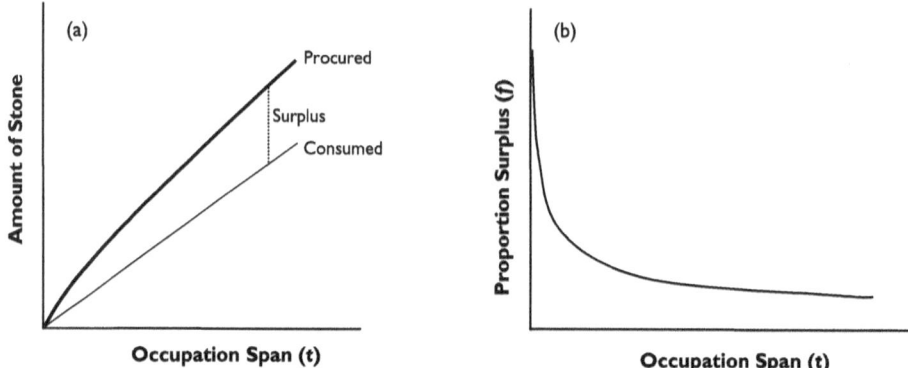

FIGURE 5.7. (a) Predicted relationships between the amounts of stone procured and consumed as a function of time. The size of the surplus (dashed line) is the difference between these amounts. (b) Generalized graph of equation 5.12, the predicted relationship between occupation span and proportion of the assemblage composed of surplus raw materials.

raw material environment is not constant, as with the five sites in the study sample. In this section, I set aside raw material effects and focus on mobility. In the next section, I return to raw material issues.

If the distribution of lithic raw materials is held constant, the model has very straightforward predictions for the archaeological record. As shown above (fig. 5.5, equation 5.11), optimum surplus size increases as a function of the square root of occupation span assuming the variable $k$ is constant. Therefore, the total raw material procured should exceed the total raw material consumed. The predicted relationship between raw material procurement and consumption is shown in figure 5.7a. The rate of procurement rises quickly as an occupation begins, but as it is prolonged the rate of procurement slows and approaches the rate of consumption. Because absolute surplus size for any given site is usually unknown, expressing surplus size as a proportion of the assemblage permits one to compare surpluses among a set of sites or components. Proportions have the added advantage of being per capita measures as well.[2] The expected proportion of lithic raw material that will be represented by surplus ($f$) is:

$$f = \frac{\sqrt{\frac{d}{e}xkt} - k}{ut + \sqrt{\frac{d}{e}xkt} - k} \qquad (5.12)$$

A graph of this equation is shown in figure 5.7b. Notice that the proportion of assemblage composed of lithic surplus declines as a function of occupation span. Although the size of the surplus should increase with longer occupations, it will not increase at a rate exceeding that of the total assemblage size. Therefore, surplus raw material becomes a smaller fraction of the assemblage.

*Puntutjarpa Rockshelter*

I return once again to Puntutjarpa rockshelter because holding place constant allows me to control for the effects of lithic raw material environment. To examine the relationship between mobility and lithic surpluses, it is necessary to divide the assemblage into those artifacts that have been "consumed" and those that constitute "surplus" raw material. A site can be provisioned with surplus raw material in the form of cores, tools, or even debitage. The fundamental distinction between surplus and consumed raw material, therefore, may not be one of form, but of size. While Gould does not provide size distributions for all artifacts for all levels, the typological scheme he employs does use size to distinguish different types of artifacts with similar forms. With respect to cores, Gould (1977:80–81) differentiates between large cores (Types 1a–c) and micro-cores (Types 1d and 1e). A similar size distinction is made with respect to flake tools and adzes. In flake tools, a distinction is made between large tools (Types 2a and b) and small tools (4a–c), and adzes (Types 3a–3d) and micro-adzes (Types 3e–3h) (Gould 1977:83–85). In the adze and micro-adze categories, Gould also differentiates between usable tools and slugs based on the size of the tool and the extent of retouch on the edge. In a more recent study of the adzes from Puntutjarpa, Hiscock and Veth (1991) have revised Gould's original typology and have shown that Gould's micro-adze slugs tend be larger than artifacts classified as micro-adzes, and suggest instead that this category of artifacts likely represents the slug stage of large adzes made on smaller flakes. No size distributions or typologies are available for debitage. Based on these considerations, the assemblage was divided into surplus and consumed raw materials (table 5.1).[3] These conceptual categories are intended to separate those artifacts that retain some use potential from those that have no potential for use. Presumably, some proportion of the assemblages is

TABLE 5.1. Typological divisions of artifact types from Puntutjarpa into surplus and consumed lithic raw materials.

| \multicolumn{2}{l}{Surplus raw material} | \multicolumn{2}{l}{Consumed raw material} |
|---|---|---|---|
| Type No. | Name | Type No. | Name |
| 1a | Large cores, 1 striking platform | 1d | Micro-cores, 1 striking platform |
| 1b | House hoof cores | 1e | Micro-cores, >1 striking platforms |
| 1c | Large cores, >1 striking platforms | 3b | Tula adze slugs |
| 2a | Large flake scrapers with retouch | 3d | Non-Tula adze slugs |
| 2b | Spokeshaves | 3e | Tula micro-adzes |
| 3a | Tula adzes | 3f | Tula micro-adze slugs |
| 3c | Non-Tula adzes | 3g | Non-Tula micro-adzes |
|  |  | 3h | Non-Tula micro-adze slugs |
|  |  | 3i | Small endscrapers |
|  |  | 4a | Lunates |
|  |  | 4b | Bondi points |
|  |  | 4c | Backed blades of irregular shape |
|  |  | 9 | Retouched frag. and utilized flakes |
|  |  | None | Debitage |

composed of artifacts that were lost (as opposed to purposeful discard), and referring to those pieces as "surplus" raw material is questionable. Since they likely represent only a very small percentage of the assemblages, I do not consider the possible inclusion of lost artifacts in the surplus category problematic. Counts of surplus versus consumed artifacts broken down by stratigraphic level are shown in table 5.2.

As predicted above, the proportion of artifacts represented by surplus raw material should decline with occupation span according to the model developed in equation 5.12 (fig. 5.7b). Because the exact relationship between the OSI (or artifact density) and occupation span is unknown, it is only predicted that a negative correlation between OSI and surplus size should be present. The data from Puntutjarpa largely confirm this prediction. Generally speaking, a trend qualitatively similar to that predicted is present at Puntutjarpa (fig. 5.8). The two deepest stratigraphic levels of the shelter (V5 and X5) show no surplus accumulation (table 5.2). The lack of surplus raw materials in these strata may be due to small sample size; fewer than one hundred artifacts were recovered from both levels combined. These strata are also characterized by the lowest OSI,

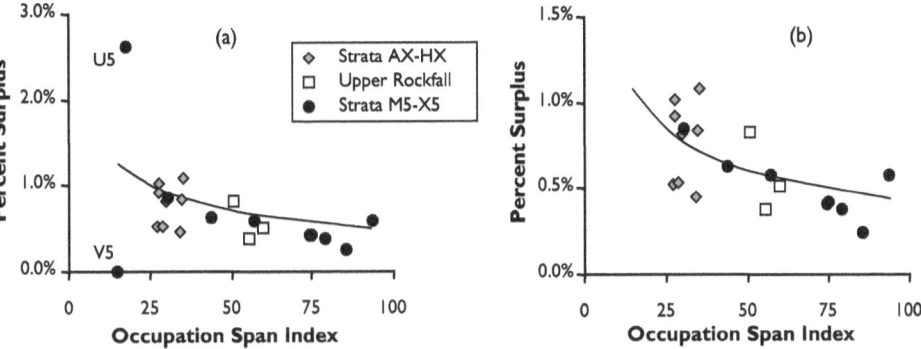

FIGURE 5.8. (a) Occupation span index vs. percent surplus raw material for the twenty-one stratigraphic levels of Puntutjarpa shelter for which the OSI can be calculated. Data labels indicate strata U5 and V5. Markers are shown by stratigraphic level. Regression line excludes stratum V5 and calculated using equation 5.12, where $dx/e = 80.987$, $k = 0.040$, and $u = 35.937$; $r^2 = 0.525$; $p = 0.0003$. (b) Same with data points for strata U5 and V5 removed. Regression line calculated using equation 5.12, where $dx/e = 17.991$, $k = 0.158$, and $u = 39.153$; $r^2 = 0.413$; $p = 0.003$.

suggesting very brief occupations and an alternative explanation—that surplus accumulation is a threshold phenomenon. If, for example, one stays at a site for only a few hours or days, there may be no incentive to accumulate a lithic surplus.

Though a significant nonlinear regression based on equation 5.12 can be fit to the Puntutjarpa point scatter (fig. 5.8), the relationship between the OSI and occupation span is unclear, so a nonparametric correlation is a more appropriate statistical test of significance. For all stratigraphic levels where surplus raw materials are present and OSI can be calculated (n = 20), there is a highly significant negative correlation between OSI and percent lithic surplus (Spearman's $\rho = -0.644$; $p = 0.002$). Therefore, as predicted, large artifacts decline in frequency with longer mean per capita occupation spans, supporting the hypothesis that lithic surplus is maintained to cushion against raw material shortfalls.

*The Study Sample*

To apply the same test to the sites in the study sample, the assemblages were divided into surplus and consumed artifacts. The division was based

TABLE 5.2. Counts of surplus and consumed artifacts from Puntutjarpa shelter.

| Level | Surplus artifacts | Consumed artifacts | Total | Percent surplus |
|---|---|---|---|---|
| AX | 4 | 766 | 770 | 0.52 |
| BX | 7 | 1,530 | 1,537 | 0.46 |
| CX | 14 | 2,608 | 2,622 | 0.53 |
| DX | 31 | 3,775 | 3,806 | 0.81 |
| EX | 42 | 4,493 | 4,535 | 0.93 |
| FX | 38 | 3,689 | 3,727 | 1.02 |
| GX | 52 | 4,740 | 4,792 | 1.09 |
| HX | 31 | 3,656 | 3,687 | 0.84 |
| IZ | 16 | 3,100 | 3,116 | 0.51 |
| JZ | 11 | 1,324 | 1,335 | 0.82 |
| KZ | 6 | 1,600 | 1,606 | 0.37 |
| LZ | 2 | 634 | 636 | 0.31 |
| M5 | 31 | 8,115 | 8,146 | 0.38 |
| N5 | 24 | 9,821 | 9,845 | 0.24 |
| O5 | 36 | 8,732 | 8,768 | 0.41 |
| P5 | 26 | 6,114 | 6,140 | 0.42 |
| Q5 | 27 | 4,630 | 4,657 | 0.58 |
| R5 | 22 | 3,768 | 3,790 | 0.58 |
| S5 | 12 | 1,896 | 1,908 | 0.62 |
| T5 | 12 | 1,399 | 1,411 | 0.85 |
| U5 | 12 | 444 | 456 | 2.63 |
| V5 | 0 | 76 | 76 | 0 |
| X5 | 0 | 18 | 18 | 0 |

on an arbitrary size cutoff of 5 cm. Artifacts having a maximum dimension larger than 5 cm were considered surplus, and those smaller than 5 cm were considered consumed.[4] Unlike Puntutjarpa, this analysis considers both artifact counts and artifact mass. Although mass and counts are expected to be collinear, the exact relationship in comparing a number of sites will depend on mean artifact sizes. Presumably, mass would be a better measure of surplus because it relates more directly to the amount of raw material present. For example, fifty flakes that are 10 cm in length would constitute a relatively smaller surplus than fifty cores that are 10 cm in length.

Table 5.3 presents artifact type counts for pieces greater than, and less than or equal to, 5 cm in length for the five sites in the study sample. As in chapter 3, this analysis is limited to artifacts larger than 2 cm to control for

TABLE 5.3. Artifact counts for the five sites of the study sample by type and maximum length.

| Artifact type | Site | | | | |
|---|---|---|---|---|---|
| | CKM | AGB | KRMP | BGB | UTM |
| Artifacts ≥ 2 cm | | | | | |
| Debitage | 17 | 383 | 500 | 1,503 | 5 |
| Flake tools | 14 | 73 | 56 | 116 | 4 |
| Bifaces | 0 | 7 | 10 | 14 | 0 |
| Channels | 5 | 46 | 9 | 11 | 0 |
| Points and preforms | 3 | 18 | 5 | 13 | 2 |
| Cores | 1 | 5† | 4 | 31 | 1 |
| Sum | 40 | 532 | 584 | 1,688 | 12 |
| Artifacts ≥ 5 cm | | | | | |
| Debitage | 1 | 13 | 11 | 62 | 1 |
| Flake tools | 3 | 21 | 7 | 23 | 1 |
| Bifaces | 0 | 7 | 6 | 10 | 0 |
| Channels | 0 | 2 | 0 | 0 | 0 |
| Points and preforms | 1 | 6 | 0 | 1 | 1 |
| Cores | 1 | 5† | 2 | 26 | 1 |
| Sum | 6 | 54 | 26 | 122 | 4 |
| % count surplus‡ | 15.00 | 10.15 | 4.45 | 7.23 | 33.33 |

*Notes:* †Only three cores were located during the analysis of the Agate Basin assemblage. This number is based on Frison's (1982b:70) count. ‡Calculated as the sum of all artifacts larger than 2 cm divided by the sum of all artifacts larger than 5 cm since the former category includes the latter.

differences in excavation and screening among sites. Comparing the OSI and artifact densities to percent surplus raw material in the assemblages (based on artifact counts), a similar distribution to that predicted by the model is evident (fig. 5.9). Based solely on variation in occupation span, Barger Gulch appears to have a slightly larger surplus than expected. The relationship between artifact density and percent count surplus (fig. 5.9b) is significant (Spearman's $\rho = -0.9$; $p = 0.037$), but because the OSI for Upper Twin Mountain (table 4.1) exceeds those of Carter/Kerr-McGee and Agate Basin, there is not a significant correlation between OSI and percent count surplus (fig. 5.9a, Spearman's $\rho = -0.6$; $p = 0.285$). Barger appears as an even greater outlier when surpluses are measured by mass (table 5.4, fig. 5.10), and there is no significant correlation between OSI

TABLE 5.4. Total artifact mass (g) for the five sites of the study sample by type and maximum length.

|  | Site | | | | |
| --- | --- | --- | --- | --- | --- |
| Artifact type | CKM | AGB | KRMP | BGB | UTM |
| Artifacts ≥ 2 cm | | | | | |
| Debitage | 33.31 | 1,043.3 | 1,160.00 | 5,041.37 | 40.21 |
| Flake tools | 156.14 | 689.3 | 383.63 | 939.03 | 23.17 |
| Bifaces | 0 | 263.49 | 241.82 | 655.85 | 0 |
| Channels | 5.07 | 59.58 | 13.12 | 19.05 | 0 |
| Points and preforms | 8.78 | 117.62 | 14.37 | 47.12 | 10.67 |
| Cores | 46.3 | 509.00† | 389.25 | 4,768.11 | 79.18 |
| Sum | 249.6 | 2,682.29 | 2,202.19 | 11,470.53 | 153.23 |
| Artifacts ≥ 5 cm | | | | | |
| Debitage | 0.45 | 257.75 | 69.47 | 1,258.48 | 24.32 |
| Flake tools | 110.68 | 431.47 | 173.88 | 571.50 | 6.46 |
| Bifaces | 0 | 263.49 | 200.46 | 619.48 | 0 |
| Channels | 0 | 7.72 | 0 | 0 | 0 |
| Points and preforms | 5.92 | 71.48 | 0 | 8.61 | 8.82 |
| Cores | 46.3 | 509.00‡ | 346.75 | 4,667.04 | 79.18 |
| Sum | 163.35 | 1,540.91 | 790.56 | 7,125.11 | 118.78 |
| % Mass surplus* | 65.44 | 57.45 | 35.90 | 62.12 | 77.52 |

*Notes*: †The mass for cores from Agate Basin was calculated using the observed mass for the three cores analyzed (305.53 g) and multiplying by 5/3 to account for the masses of the two missing cores. ‡It was assumed that both missing cores were larger than 5 cm in maximum dimension. *Calculated as the sum of the masses of all artifacts larger than 2 cm divided by the sum of the masses of all artifacts larger than 5 cm since the former category includes the latter.

or artifact density and percent mass surplus. If a regression line, based on equation 5.12, is fit to the four other sites (excluding Barger) using artifact densities as a proxy for occupation span, Barger Gulch is predicted to have an assemblage composed of approximately 25.3 percent surplus, but it deviates significantly from that value at 62.1 percent. This is not necessarily problematic. Because site-specific raw material distributions will affect the relative costs of direct and embedded procurement, a straightforward relationship between occupation span and surplus size might not be expected. It is somewhat surprising that the variability seen among the other four sites is largely explainable using only differences in occupation span.

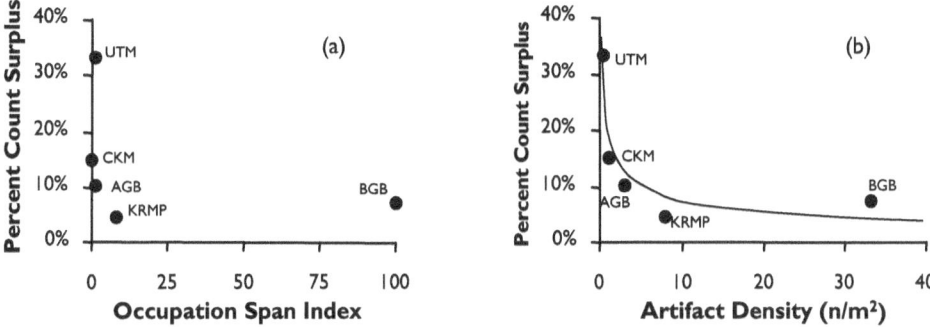

FIGURE 5.9. (a) Occupation span index vs. percent surplus based on artifact counts. (b) Artifact density vs. percent count surplus. Regression line calculated using equation 5.12, where $dx/e = 4.624$, $k = 1.46 \times 10^{-5}$, and $u = .032$; $r^2 = 0.858$; $p = 0.024$.

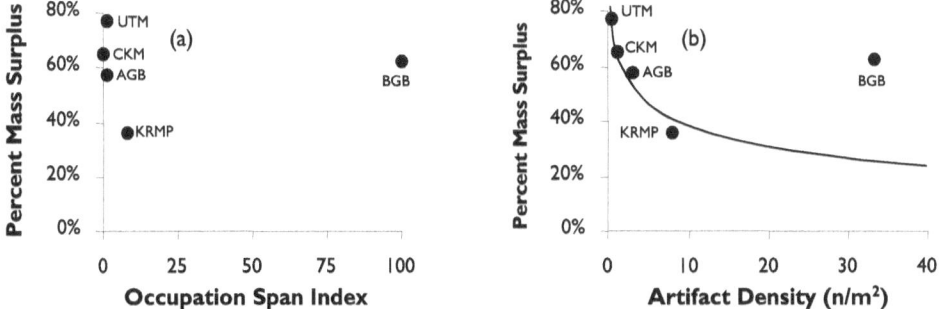

FIGURE 5.10. (a) Occupation span index vs. percent surplus based on artifact mass. (b) Artifact density vs. percent mass surplus. Regression line calculated using equation 5.12 excluding Barger Gulch, where $dx/e = 5.485$, $k = 0.0001$, and $u = .012$; $r^2 = 0.963$; $p = 0.019$.

Also of interest when considering the size of the surpluses with respect to mass is that every site, with the single exception of Krmpotich, has more surplus mass than consumed raw material. However, this analysis considers only artifacts larger than 2 cm, and if the masses of the entire assemblages from the sites were used instead, these percentages would drop considerably. Additionally, it is expected that some proportion of the artifacts that were used at a particular site would not be discarded there. These factors artificially inflate the relative proportions of surplus raw material within the assemblages.

## On the Relative Costs of Direct and Embedded Procurement

The model predicts that surplus size should increase as a square root function of the ratio of the costs of direct to embedded procurement. To explore the effects of raw material availability, it is necessary to consider what controls the relative costs of direct and embedded procurement. The cost of direct procurement should primarily be a function of distance. Other factors such as terrain and load will also come into play, but in the simplest scenario the greater the distance to a raw material source, the greater the cost of procurement. Does this relationship hold true for embedded procurement as well? Binford (1979:259) notes that embedded procurement comes with "few or no direct costs accountable for the procurement of raw materials used in the manufacture of implements." Whether procurement is direct or embedded, however, minimally the same distance trip must be made and the same load carried. These are direct costs resulting from the procurement of raw materials, and if procurement is embedded, the trip may be longer, both in terms of time and distance because the trip may not be direct.

Consider the following example. You need to do two things today. You need to go to work and pick up some asparagus for dinner. If you can stop by the grocery store on the way home from work, you will save more time and gas than if you were to make two trips, one to work and one to the grocery store. If the grocery store is not located on the fastest, most direct route from work to home, then you must make a detour from your normal commute. This is a real cost, but in most situations the added length of the detour will still be less than the direct trip from your home to the grocery store. Obviously, the cost savings of embedded asparagus procurement depends on the locations of home, work, and the store.

Returning to mineral after a brief vegetable detour, while it is generally true that embedded procurement will be less expensive in terms of time, energy, and opportunity costs than direct procurement, the precise ratio of costs depends on local raw material environments. In this section, I build a simple computer simulation to explore how the distribution of lithic raw material sources relative to a central place affects the $d{:}e$ ratio. In the simplest simulation, a single point source of raw material is placed at varying distances. When the forager engages in direct procurement,

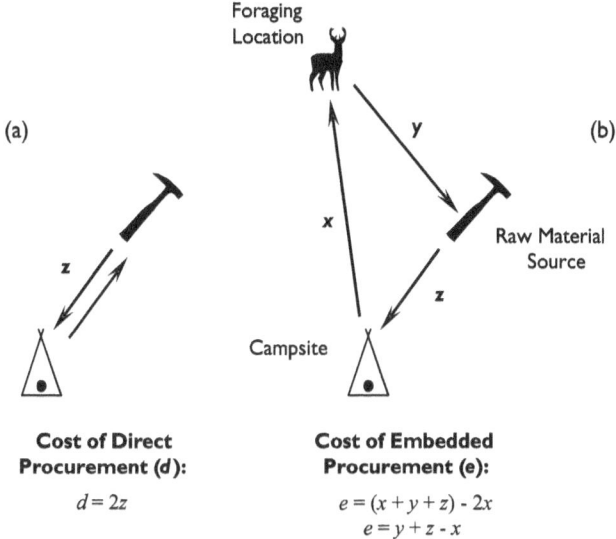

FIGURE 5.11. A simple model of raw material procurement wherein (a) the cost of direct procurement is calculated as the round-trip distance between the campsite and the raw material source, and (b) the cost of embedded procurement is calculated as the added distance walked in the raw material procurement portion of the trip, or the sum of all distances in the trip minus the round-trip foraging distance.

she makes a round-trip foray in a straight line from a central location to the raw material source and back again. The cost of direct procurement ($d$) is calculated as the round-trip distance from the central location to the raw material source (fig. 5.11a). I use distance traveled in raw material procurement as the measure of cost and do not include opportunity costs, though they should be directly proportional to resource costs. When the forager engages in embedded procurement, a foraging location is randomly chosen, defined by a distance and bearing (random numbers drawn from uniform distributions). The forager travels in a triangular pathway from the central place to the foraging locale, from the foraging locale to the raw material source, and from the raw material source back to the campsite. The cost of embedded procurement ($e$) is calculated as the added distance involved in the portion of the trip devoted to raw material procurement (fig. 5.11b). Specifically, it is the difference between the total distance traveled and the round-trip foraging distance had the additional lithic procurement stop not occurred. For the one-source simulation,

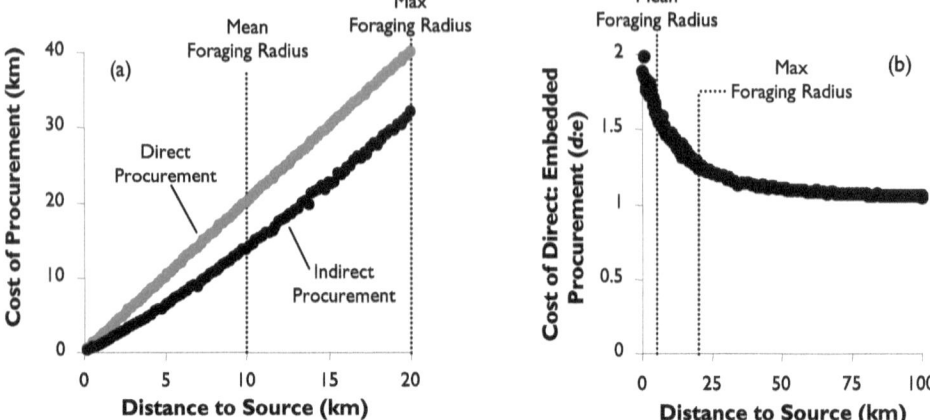

FIGURE 5.12. (a) Costs of direct and embedded procurement (estimated by simulation) for 1,000 logistical forays for each raw material distance vs. distance to a single source of lithic raw material assuming a maximum foraging radius of 20 km. (b) The relative costs of direct:embedded procurement ($d$:$e$) vs. distance to a single source of raw material under the same assumptions.

a single raw material source was placed at distances varying from 0.1 to 100 km from a central location. For each run of the simulation, foraging locations (n = 1,000) were chosen based on random distances (0 to 20 km) and bearings (0 to 360°). The cost of embedded procurement for each foray was calculated and averaged for the one thousand logistical trips.

Figure 5.12 shows the relationship between the distance of the source from the central place and the relative costs of direct and embedded procurement. While the cost of direct procurement increases as a linear function of distance to source, the relationship between distance and cost of embedded procurement is not linear (fig. 5.12a). Thus, the $d$:$e$ ratio is affected by distance to source (fig. 5.12b). For this model, the $d$:$e$ ratio ranges from 1 to 2, and somewhat counterintuitively is greatest when lithic raw material sources are nearby. Perhaps the simplest way to understand this phenomenon is to consider a lithic raw material source that sits 1,000 km away from a campsite where the inhabitants regularly engage in foraging trips of up 20 km one way. If the forager moves in the exact direction of the raw material source, the cost of embedded procurement

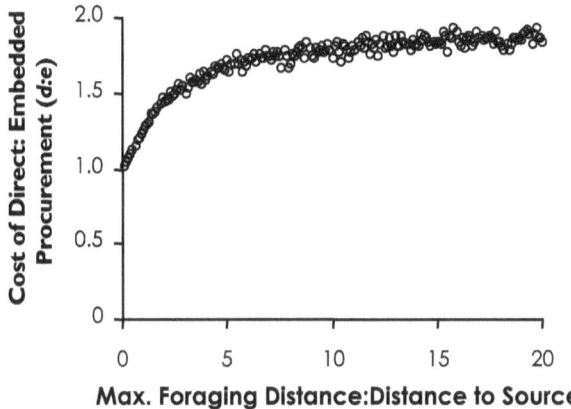

FIGURE 5.13. The ratio of maximum foraging distance to distance to a single raw material source vs. the *d:e* ratio, estimated by simulation for 1,000 logistical forays at each foraging distance.

(as calculated in fig. 5.11) would be 1,960 km and the cost of direct procurement would be 2,000 km, a negligible difference. If the source is moved closer and closer to camp, the relative difference increases. When the source sits very close to camp, both direct and embedded procurement are inexpensive, but because the source will never fall far from any logistical foray, embedded procurement is at its cheapest. From this example, it should also be apparent that a second critical variable is the average distance traversed in logistical forays.

A second single-source simulation was performed where the distance to source was held constant at 5 km, and maximum foraging distances were varied from 0.5 to 100 km. The results of this simulation are shown in figure 5.13. As the foraging distances increase relative to the distance of a raw material source, the *d:e* ratio increases toward a maximum value of 2. Therefore, for a single raw material source, direct procurement is most expensive relative to embedded procurement when the source is close to camp. The word "close" must not be used in an absolute sense but must be considered relative to typical foraging distances.

The results for the single-source simulation are fairly straightforward, but things get somewhat more complex when multiple sources are added. In the multiple-source simulation, foraging locations are chosen from a constant uniform random distribution varying between 0 and 20 km, and bearings again are likewise randomly drawn (0 to 360°).

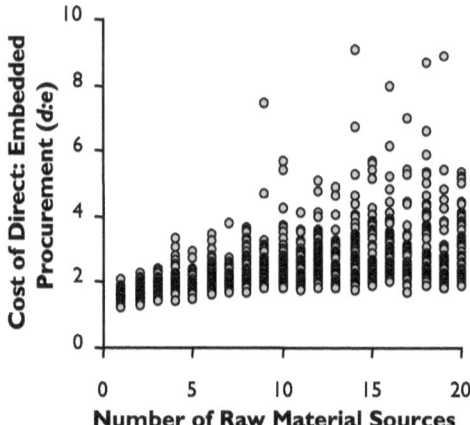

FIGURE 5.14. The number of lithic raw material sources (randomly placed) vs. the *d:e* ratio for multiple-source raw material configurations.

Multiple sources ($2 \leq n \leq 20$) are randomly placed on the landscape the same way foraging locations are drawn. For each multiple-source configuration, the cost of direct procurement is calculated as the round-trip distance between the central place and the closest raw material source. For each logistical foray, minimum distances are calculated between each raw material source and the line segment defined by the central place and the foraging location. The raw material source closest to the line segment is visited and its position is used to calculate the cost of embedded procurement. Therefore, for any given multiple raw material source configuration, embedded procurement may occur at different raw material sources during different logistical forays. For each multiple source configuration, the cost of embedded procurement is averaged for one hundred logistical forays.[5]

Figure 5.14 shows the relationship between the number of sources and relative cost of direct to embedded procurement for 1,069 runs of the simulation. Generally speaking, as the number of available sources increases, so does the *d:e* ratio because the average length of detours necessary to embed raw material procurement within foraging decreases. Thus, the average cost of embedded procurement declines. With greater numbers of sources, the degree of dispersion in the *d:e* ratio appears to increase, meaning the number of available sources by itself is not a good predictor of the *d:e* ratio. In the multisource simulations, values of *d:e* as

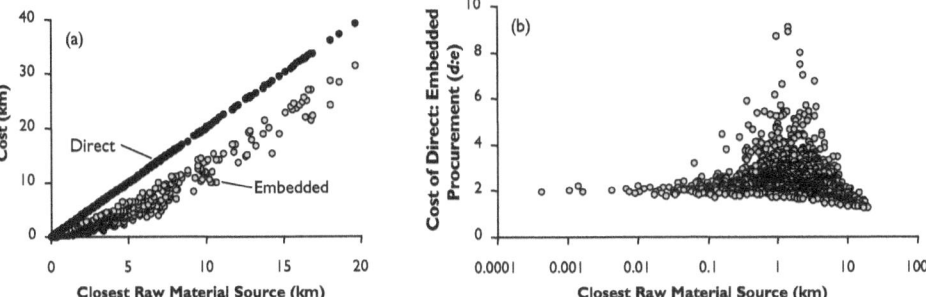

FIGURE 5.15. (a) The closest raw material source vs. the cost of direct and embedded procurement for multiple-source (randomly located) raw material configurations assuming random foraging up to 20 km in distance. (b) The closest raw material source vs. the $d{:}e$ ratio for randomly placed raw material source configurations.

high as 9.06 were observed, and with more sources the number could be even higher.

A second major factor affecting the $d{:}e$ ratio in the multiple-source simulation is the distance to the closest available raw material source, because it determines the cost of direct procurement and strongly affects the cost of embedded procurement (fig. 5.15a). The $d{:}e$ ratio appears to be maximized when the closest source sits ≈0.6 km from the central place and a large number of sources (n ≥9) are available (fig. 5.15b). This distance, however, is dependent on foraging distances. For the multisource simulations a maximum foraging distance of 20 km and a mean foraging distance of 10 km were assumed. When the closest source sat at distances greater or lesser than ≈1.6 km from the central place, the maximum observed $d{:}e$ ratio declined. This finding differs from the single-source simulation where $d{:}e$ increased continuously as the source was moved closer (fig. 5.12). A more systematic analysis of the simulation would certainly clarify these relationships.[6]

## Calibrating Barger Gulch

Barger Gulch is well known for an abundance of locally available lithic raw materials (Izett 1968; Kornfeld and Frison 2000; Kornfeld et al. 2001; Naze 1986; Surovell et al. 2001; Waguespack et al. 2002; White 1999). The site sits amid the chert-bearing Miocene Troublesome Formation,

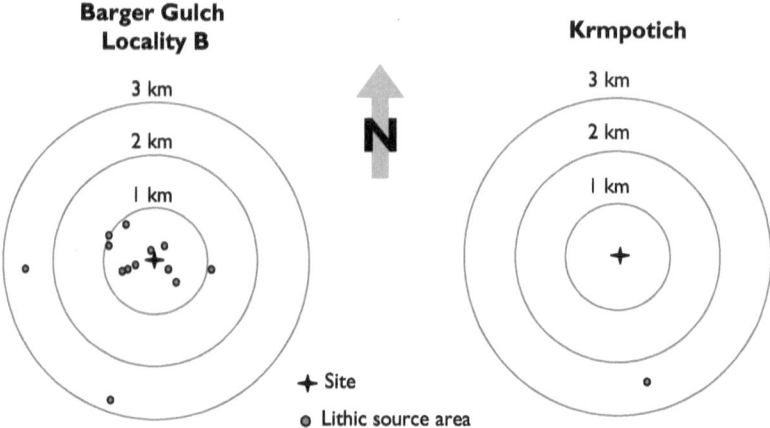

FIGURE 5.16. Known raw material source areas within 3 km of the Barger Gulch, Locality B (left) and Krmpotich (right) sites.

and outcrops of high-quality chert are available in virtually any direction from the site within 2 km distance. It would be a rare event to walk 1 km from the site and not encounter a nodule of chert. In this respect, embedding raw material procurement within other activities should be extremely efficient.

How does Barger Gulch compare, in this respect, to the other four sites in the study sample? All of the sites have lithic raw materials available within a 5-km distance, but none approach the density and ubiquity of raw materials surrounding Barger Gulch, Locality B. Figure 5.16 shows, for example, a comparison of mapped raw material outcrops within a 3-km distance of the Barger Gulch and Krmpotich sites. At Barger Gulch numerous outcrops of Troublesome Formation chert are available within a 1-km radius, while Krmpotich has only a single source of oolitic (i.e., ostracodal) chert available at approximately 2.5 km to the south. It should be noted that numerous other raw material source areas are present within 20 km of both sites. Not shown in figure 5.16, for example, is the deeply incised dendritic drainage pattern of Barger Gulch, which serves to concentrate nodules of chert eroded from the Troublesome Formation. Barger Gulch proper comes within 300 m of the site to the west and 700 m to the south, and could have served as another source area for the site occupants. Although raw material surveys for Barger Gulch and Krmpotich are not complete, additional survey data would likely increase the great discrepancy in raw material distributions seen between the two sites.

The raw material environments for the remaining three sites are not as well known. At the Agate Basin site, moss agate and petrified wood are available within a 1-km distance, and Frison describes their distribution as "scattered" (Frison 1982a:178). Personal experience at the site confirms this description since large workable nodules of this material are seldom found, at least today. Therefore, although the distance to raw material at Agate Basin is similar to that of Barger, the probability of locating workable nodules of lithic raw material is significantly lower. At CKM, local raw materials include porcellanite and petrified wood. Reiss et al. (1980:27–28) report primary outcrops and secondary sources of porcellanite "within the site area," and petrified wood is reported to be "known throughout eastern Wyoming, including the eastern Powder River Basin." Although it is difficult to discern a clear picture of it, the raw material distribution at CKM is likely similar to that of Agate Basin, with a low density of scattered nodules within reach of the site. Like Barger Gulch, the Upper Twin Mountain site sits atop the Troublesome Formation, but unlike Barger, it is characterized by a significantly lower density of chert with small scattered nodules available within 1 km of the site and more extensive outcrops available within 5 km (Kornfeld, personal communication, 2002). Also available within 1 km of the site is a low-quality but flakeable granodiorite that is present in the assemblage. Compared to other sites, the relatively large surplus at Barger Gulch, Locality B, may be due to the extremely low cost of embedded procurement relative to direct procurement at the site.

It is possible to use the simulation coupled with the known raw material distributions for Barger Gulch and Krmpotich (fig. 5.16) to actually estimate $d$:$e$ ratios for the two sites, permitting approximate prediction of the effects of raw material environments on surplus size for each location, respectively. To do this, the positions of the mapped raw material sources relative to each site are plugged into the simulation. Figure 5.17 shows the estimated $d$:$e$ ratios for the two sites assuming random foraging at various distances. Although differences in foraging distance between the two sites does affect the $d$:$e$ ratios, because raw material sources sit nearby both sites, the ratios quickly stabilize above maximum foraging distances of approximately 5 km. Therefore, the effect of differences in foraging distance for the two sites may be negligible. Averaging estimated values of $d$:$e$

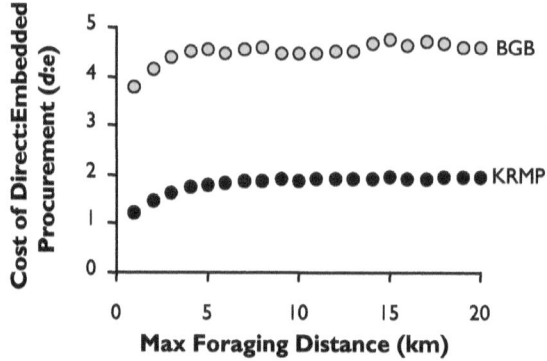

FIGURE 5.17. Maximum foraging distance vs. the cost of direct:embedded procurement, estimated by simulation for Barger Gulch, Locality B (gray circles) and Krmpotich (black circles) based on the known raw material sources within a 3-km distance from each site (fig. 5.16).

for all runs where maximum foraging distance is ≥5 km, for Barger Gulch, the estimated *d:e* ratio is approximately 4.54, and for Krmpotich it is 1.89.

Plugging these values into the *d:e* ratio in equation 5.11, the surplus size for Barger Gulch can be adjusted to an expected value assuming it had a raw material environment like that of the Krmpotich site. I use the observed artifact density (33.1 artifacts/m²) as a proxy for occupation span ($t$). Assuming that $k$ is equal to a value between 0.00001 and 1.0, and $x$ ranges between 200 and 1,000 g,[7] the optimal surplus size for Barger Gulch would be predicted to be reduced by 64.3 to 64.5 percent in a Krmpotich-like raw material environment. To calculate adjusted surplus sizes, I multiplied to observed absolute count- and mass-based surplus sizes (tables 5.3 and 5.4) by 64.4 percent, and recalculated percent surplus. When this is done, Barger has an adjusted percent count surplus of 4.8 percent and an adjusted percent mass surplus of 51.4 percent. As shown in figure 5.18, even after controlling for raw material distributions the percent mass surplus at Barger Gulch is still larger than expected, but now the correlations between both artifact density and percent mass and count surpluses are significant (for both correlations: Spearman's $\rho = -0.9$; $p = 0.037$). It is possible and perhaps likely that with a more complete raw material survey at Barger Gulch that the *d:e* ratio would be increased further, and that the adjusted surplus size would be in turn reduced further.

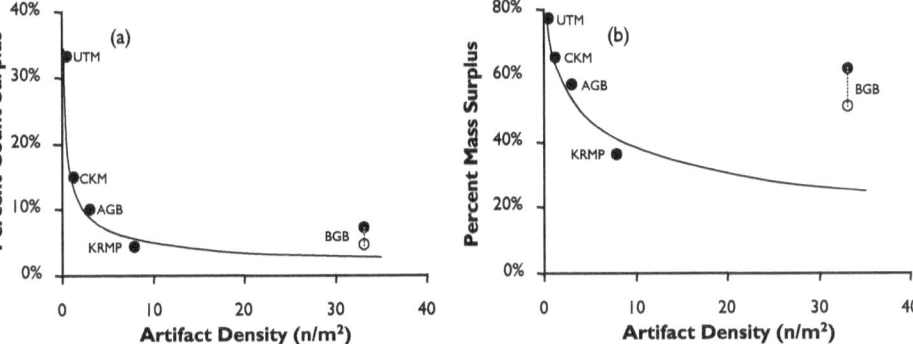

FIGURE 5.18. (a) Artifact density vs. percent count surplus. White circle indicates simulation-adjusted surplus size for Barger Gulch. Regression line based on equation 5.12 excluding Barger Gulch, where $dx/e = 300.12$, $k = 0.003$, and $u = 5.654$; $r^2 = 0.957$; p = 0.022. (b) Artifact density vs. percent mass surplus. White circle indicates simulation-adjusted surplus size for Barger Gulch. Regression line calculated using equation 5.12 excluding Barger Gulch, where $dx/e = 5.485$, $k = 0.0001$, and $u = .012$; $r^2 = 0.963$; p = 0.019.

## Discussion

In a classic paper dealing with large-scale patterns in the archaeological record, Parry and Kelly (1987) explore the relative trade-offs between formal and "expedient," or unsystematic, core and tool technologies. They argue that formal core reduction (e.g., bifacial or blade technology) is advantageous to mobile hunter-gatherers because it minimizes production waste, therefore maximizing portability. The downside of formal reduction is the relatively high cost of production, use, and maintenance. In contrast, expedient production is not particularly efficient in terms of raw material use and portability but has significantly lower production costs. They argue further that the unmodified edges of informal tools may be more efficient in use than the retouched edges of formal tools. The switch in late prehistory in many parts of the globe from formal core reduction to informal rock knocking is then seen as a correlate of decreasing mobility. Sedentism, they suggest, comes with the fringe benefit of the ability to stockpile lithic raw materials at residential locations, producing an ultra-local lithic raw material source, which in turn would eliminate the advantage of producing core and tool forms designed for portability.

Maintaining a stockpile of raw materials, of course, means that when a location is abandoned that stockpile will be abandoned as well, no matter how much energy is invested into the accumulation of that surplus. From the perspective of Parry and Kelly (1987), the benefit of producing a lithic surplus, then, is to reduce the costs of tool production and increase task efficiency through the production of more efficient edges.

In a similar model, Kuhn (1995:24–31) explores the two alternative raw material provisioning strategies—the provisioning of individuals and the provisioning of places. In this model, highly mobile hunter-gatherers should emphasize the provisioning of individuals because of unpredictability in the congruence between tasks and lithic raw materials. Because lithic raw materials are not ubiquitous, the best way to avoid lithic shortfalls when highly residentially or logistically mobile is to bring your stone with you. For less mobile groups with relatively long occupation spans, the emphasis instead should be on provisioning a place with lithic raw material to ensure a ready supply of tool stone when gearing up for logistical forays. Therefore, it is expected that very short-term occupations will exhibit little evidence of provisioning of campsites, whereas long-term occupations would show extensive provisioning. Following Parry and Kelly (1987), Kuhn (1995:27–28) sees the use of informal technologies as evidence of the provisioning of places, but he also suggests that the relative quantities of expedient and formal tools in archaeological sites will not necessarily indicate the relative dependence on the two alternative strategies, because expedient and formal tools have dramatically different life histories.

Another plausible explanation of the apparent abandonment of surplus raw material is caching. Stockpiles are not purposely abandoned because the individuals who created them intended to return to them at a later date. Therefore, the initial investment in procurement effort is made with the intent that it would pay off at some point in the future. Artifact concentrations that have been referred to as caches are certainly known from Paleoindian times (e.g., Butler and Fitzwater 1965; Collins 1999; Frison and Bradley 1999; Gramly 1993; Green 1963; Hartwell 1995; Morrow and Morrow 2002; Stanford and Jodry 1988; Wilke et al. 1991), but they take on a very different form from the assemblages of the study sample. Others have used caching to explain the presence of large or unbroken artifacts in Paleoindian campsite contexts in relatively raw material–poor areas (Ballenger 1999; Hofman et al. 1990).

The model developed in this paper, which I have coined the "Rainy Day" model, comes to a conclusion similar to those of both the Parry and Kelly (1987) and the Kuhn (1995) studies. Lithic stockpiling and the provisioning of places are expected to correlate with occupation span. Unlike in the Parry and Kelly (1987) model, however, the advantage of lithic surplus accumulation is seen as a means of preventing periodic shortfalls of lithic raw material that would demand supplementary direct procurement forays. Caching of lithic raw material is certainly an alternative explanation, and I would hesitate to extend the hypothesis of surplus creation to prevent lithic deficits to all Paleoindian sites. However, the Rainy Day model provides one explanation for stockpiling that does not require an assumption of a temporal scale of payoff that exceeds the occupation of a site. In other words, it is assumed that the accumulated surplus was created with no intent of returning to it.

The model predicts that larger surpluses are expected with longer occupations, because the probability of suffering lithic budget deficits is greater, and because the cost of surplus accumulation is lessened by spreading the cost over an extended period. Furthermore, larger surpluses are expected as the cost of embedded procurement decreases relative to the cost of direct procurement. Support for both predictions is seen in the two data sets explored.

Employing artifact data from Puntutjarpa allowed the investigation of mobility effects by holding natural raw material distributions constant, and the proportion of the assemblages composed of surplus per stratigraphic level fits well with the predicted relationship based on occupation span. Folsom and Goshen assemblages from the study sample also fit well with the predicted mobility relationship, with the single exception of Barger Gulch, Locality B, which shows a significantly larger surplus than expected. Based on a simple computer simulation, it was argued that the proximity and density of Troublesome Formation chert in the vicinity of the site significantly lowered the cost of embedded procurement relative to direct procurement, allowing large surpluses to be created at relatively low cost.

There are alternative explanations of the large surplus seen at Barger. If Barger, as suggested in chapter 3, represents a winter occupation, then the possibility of winter snowfall could have inspired the inhabitants of the site to accumulate large surpluses in the event that raw material procurement would be hampered by snow cover and poor visibility.

This type of surplus is akin to food storage where certain subsistence resources are seasonally unavailable and stored foods serve the purpose of bridging a temporal gap. However, this type of lithic surplus accumulation is not at odds with the model. In fact, the model predicts it. The proportionality constant $k$ used in the model describes the relationship between the size of a surplus and the probability of a shortfall. Although to this point I have explored only variability in demand, if the supply of lithic raw materials is very unpredictable, as we might expect in a region where snowfall can obscure lithic raw materials from view, then $k$ increases and larger surpluses are expected. As shown in figure 5.6, larger surpluses are expected with larger values of $k$ until $k$ becomes so large that surplus accumulation does little to prevent shortfalls. The abundance of raw material at Barger is unquestionable, but the seasonality of raw material procurement may also explain the relative size of the surplus.

How would such a model be operationalized for hunter-gatherers going about their day-to-day tasks? Binford (1979:259) notes that the Inuit have a saying that "only a fool comes home empty handed!" I imagine this ethos is common among foraging populations. The statement, as Binford notes, suggests that the Inuit always keep a supply of raw materials on hand and do so by bringing home extra raw materials when they do not "have their hands full," literally speaking. Returning to camp with a nodule of lithic raw material when it is encountered during foraging may, depending on natural raw material distributions, ensure the creation of lithic stockpiles, with little or no opportunity cost. In this way, such cultural rules may codify complex mathematical formulations of human behavior in very simple terms and ensure optimality in procurement behavior.

Because I began this chapter with politics, I would be remiss not to return to it. One obvious objection a reader might have is that my Rainy Day model does not explain the behavior of politicians. Could it be that prehistoric hunter-gatherers behave optimally, but twenty-first-century politicians do not? Quite the contrary, I would argue, and the issue is really about what the term "optimal" means in the framework of behavioral ecology. Optimal means making the correct decision given a specific set of assumptions concerning constraints, decision variables, currencies, and goals. What is optimal for one may not be optimal for another. Most modern politicians (though many would likely disagree) make decisions based on the goal of maximizing the democratic currency of political popularity. Therefore, issuing huge tax refunds in the face of massive federal debt may in fact be optimal. The outcome of the 2000 presidential election would suggest as much. Furthermore, lithic deficits incurred by prehistoric hunter-gatherers had immediate consequences. By and large, budget deficits today have delayed costs that can be passed on to future politicians and taxpayers. Optimizing short-sighted political goals while sacrificing long-term fiscal health may not be sound economic policy, but it sure sounds good in a presidential debate.

# 6

# Bifaces, and So On

MODELING THE DESIGN OF TOOLS AND TOOLKITS

DESIGN ISSUES HAVE been at the forefront of lithic studies since the emergence of the paradigm known broadly as "technological organization." In this framework, alternative technological strategies are seen as ways of meeting tool stone needs as they play out within differing environmental contexts. This generalization would apply to the well-known dichotomous classifications of technology as curated or expedient (Binford 1973, 1977), maintainable or reliable (Bleed 1986), or versatile or flexible (Shott 1986). I do not mean to imply here that these typologies necessarily define mutually exclusive characters of technological entities, although they are often portrayed as such. A tool or toolkit, for example, could certainly be reliable and maintainable (Eerkens 1998). I wish only to point out that these frameworks, by which we commonly attempt to monitor variability in lithic assemblages, are attempting to define optimizing technological decisions made within an environmental setting.

I share the same goal but approach the question of tool and toolkit design through the use of formal mathematical models, in large part derived from Kuhn's (1994) model of the design of mobile toolkits. In this chapter, I explore the constraints and currencies governing the design and use of bifacial and nonbifacial technologies by Folsom and Goshen hunter-gatherers. These analyses lead to a description of the roles of flake tools, bifaces, and cores in Paleoindian toolkits. In the following chapter, I explore the factors governing intersite variability with respect to the discard of debitage produced from bifaces and cores.

I would argue that there has been a love affair between Paleoindian archaeologists and bifaces, the remaining aspects of Paleoindian technology receiving proportionately so little attention that one might wonder if we even regularly analyze them. This chapter will demonstrate that although bifacial reduction and its many benefits were commonly exploited by the prehistoric subjects of this study, the less heralded reduction strategy of "opportunistic flake production" (Frison and Bradley 1980:18) was of at least equal importance in Folsom and Goshen toolkits.

## The Kuhn Model of Mobile Toolkits

As introduced in chapter 1, Kuhn (1994) developed a formal model exploring technological trade-offs in mobile toolkits. Mobile toolkits are "artifacts that mobile individuals keep with them most or all of the time, implements that are subject to virtually continuous transport" (Kuhn 1994:427). The model is used to ask whether a forager, attempting to maximize toolkit utility per unit mass, should transport cores or flake tools/blanks. The model also predicts the optimal size at which to manufacture unifacial tool blanks assuming a goal of maximizing utility per unit mass. The following discussion, then, is taken from Kuhn (1994), although I modify the original notation and use analytical instead of graphic solutions.

### The Design of Flake Tools

The Kuhn model uses a handful of simplifying assumptions concerning how transport costs and potential utility are measured for tools versus cores. It is reasonably assumed that the cost of transporting artifacts is proportional to artifact volume (or mass). The cost of transporting a flake tool ($c_t$) is calculated as:

$$c_t = fl_t^3 \tag{6.1}$$

where $l_t$ is the length of a tool and $f$ is a constant. Since length, width, and thickness are allometrically related for lithic debitage, the dimensions of flake tools (length, width, and thickness) can be expressed as a power function of length. The constant $f$ is the ratio of thickness to length. Flakes that are thin relative to their length and width (e.g., soft hammer percussion flakes) will be characterized by relatively small values of $f$.

The utility of flake tools is measured as the amount of usable edge that can be produced from a blank through resharpening, minus the length of the unusable portion, or slug (fig. 6.1). Kuhn uses two equations to estimate the utility of flake tools ($u_t$). The first equation is a linear measure of utility:

$$u_t = l_t - m_t \tag{6.2}$$

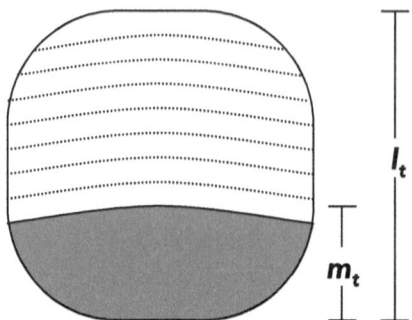

FIGURE 6.1. Schematic representation of the modeled flake tool. The shaded portion of the tool represents the unusable portion, and the unshaded portion represents the usable portion. Dashed lines represent edge positions after successive episodes of resharpening (adapted from Kuhn 1994:fig. 1).

where $m_t$ is the size of the tool slug. The second measure assumes that utility ($u_t$) is proportional to the area of the usable portion of the stone tool:

$$u_t = l_t(l_t - m_t) \tag{6.3}$$

The second equation is probably more realistic since it takes the width of the tool edge into account. Taking the ratios of utility to transport cost provide measures of flake tool transport efficiency ($e_t$):

Linear Utility        Areal Utility

$$e_t = \frac{l_t - m_t}{fl_t^3} \qquad e_t = \frac{l_t(l_t - m_t)}{fl_t^3} \tag{6.4}$$

As shown in figure 6.2, transport efficiency does not uniformly increase as a function of length. In fact, for each equation there is an optimal length that maximizes tool utility per unit mass. To calculate the tool blank length that maximizes transport efficiency, the first derivative of each equation is solved with respect to length:

Linear Utility        Areal Utility

$$\frac{de_t}{dl_t} = \frac{3m_t - 2l_t}{fl_t^4} \qquad \frac{de_t}{dl_t} = \frac{2m_t - l_t}{fl_t^3} \tag{6.5}$$

Setting each equation equal to zero and solving for length ($l$) provides the optimal solutions ($l_{opt}$):

Linear Utility        Areal Utility

$$l_{opt} = \frac{3}{2}m_t \qquad l_{opt} = 2m_t \tag{6.6}$$

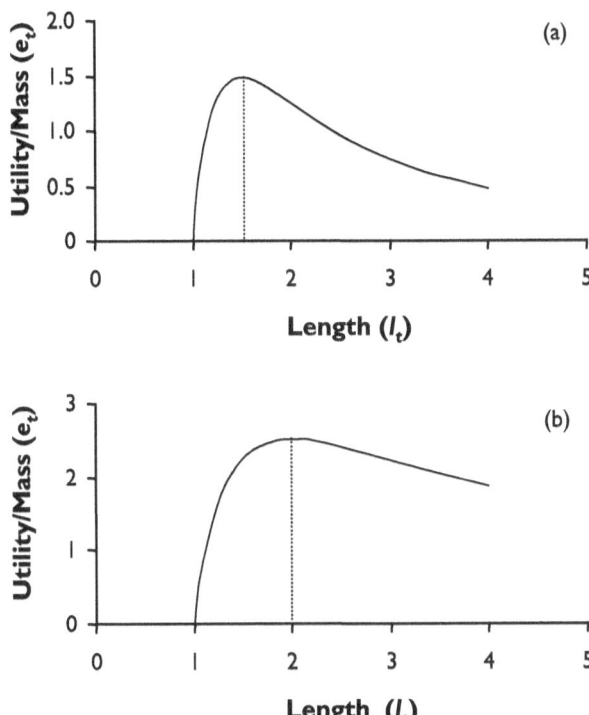

FIGURE 6.2. (a) Flake tool length ($l_t$) vs. transport efficiency ($e_t$) assuming a linear measure of utility. (b) Same, assuming an areal measure of utility. Both examples assume $m_t = 1$ and $f = 0.1$.

The model, therefore, predicts that tools should be either 1.5 their minimum usable size if utility is measured linearly, or 2.0 times their minimum size if an areal measure of utility is used. The latter prediction differs from that of Kuhn (1994:435), who suggests that the model predicts that flake tools should be optimally designed to be between 1.5 and 3.0 times their minimum usable size. The reason for this discrepancy is unclear, but both predictions cannot be correct.

The model can also be used to predict the optimum thickness:length ($f$) for flake tools. Because the utility of tools is measured one- or two-dimensionally, and the transport cost of tools is measured three-dimensionally, it is expected that tool blanks should be chosen to be as thin as possible to minimize their weight, and from equation 6.4 and figure 6.3 this expectation is born out. Larger values of $f$ (thick flakes relative to length) equate to reduced transport efficiency. Therefore, although there

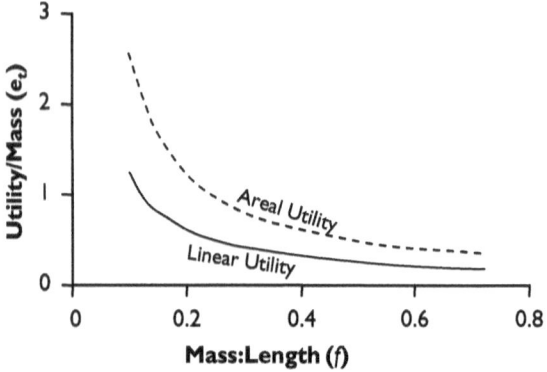

FIGURE 6.3. The ratio of flake thickness:length ($f$) vs. transport efficiency ($e_t$) assuming $l_t = 2$ and $m_t = 1$.

is no optimum, per se, the optimal solution to maximizing transport efficiency is to make tools as thin as possible relative to their length.

## Testing the Flake Tool Design Model

The Kuhn model of flake tool design predicts that upon manufacture: (1) tools should be 1.5 to 2.0 times their minimum usable size, and (2) stone tool users will prefer blanks that are especially thin relative to their length and width. In this section, I test both of these predictions. Both tests are problematic since flake tools recovered archaeologically do not preserve the original dimensions of the blank, which are required for a true test of the model. Nonetheless, it is possible to roughly explore these predictions without such information.

The first prediction comes with a relatively straightforward expectation for the size of discarded tools. Although flake tools can be discarded at any stage in their use-lives, if they do not begin larger than two times their original length, then no flake tool, holding type (and therefore minimum usable size) constant, should be greater than 2.0 times the length of the smallest example recovered. Here I limit my analysis to endscrapers because they best match the prototype of the modeled tool (fig. 6.1), and they are widely agreed to be serially reduced flake tools. Analysis of endscrapers is limited to pieces where artifact length at discard is known, with length measured as the distance from the platform to the distal tool edge, parallel to the axis of percussion.

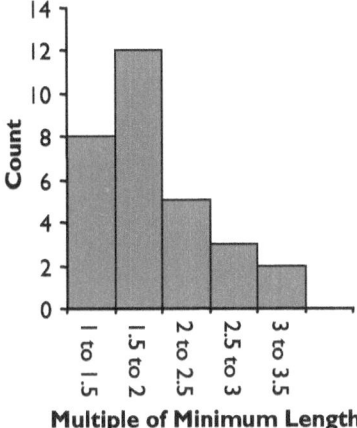

FIGURE 6.4. Histogram of endscraper lengths (n = 30), expressed as a multiple of the shortest scraper (19.44 mm).

From the five sites in the study sample, thirty-nine endscrapers were analyzed. Of those, reliable length measurements (platform to distal unbroken edge) could be made on thirty artifacts. Endscraper lengths range from 19.4 mm to 64.7 mm. The distribution of endscraper lengths, expressed as a multiple of the lowest value (19.4 mm), is shown in figure 6.4. The data do not provide support for the model in that the largest endscrapers are over three times the length of the smallest, the largest artifact being roughly 3.3 times the length of the smallest. One objection to this analysis could be that it not only includes tools transported between sites, constituents of a mobile toolkit, but also tools made and used locally, essentially constituents of an "immobile toolkit." If tools on local raw materials are excluded, the largest scraper is still 3.08 times the length of the smallest. There are numerous explanations for the discrepancy between model and data; most obvious perhaps is that some other currency is more important in decisions concerning scraper design, such as functional efficiency. For example, it is possible that the work efficiency of scrapers increases with respect to size (see Kuhn 1994:fig. 8).

The second prediction is that retouched flake tools should be made as thin as possible to minimize weight, thus maximizing transport efficiency. Although no optimum can be defined analytically, the thinness of flake tools relative to length is obviously constrained by fracture mechanics. This variable can be controlled intentionally by knapping technique.

For example, it is well known that soft hammer and bifacial reduction on average produce flakes that have lower thickness:length and thickness:width ratios than hard hammer and amorphous core reduction (Amick et al. 1988:29; Prasciunas 2004). It is difficult to measure these ratios directly for flake tools because retouch modifies the length and width of the original flake blank. Some characters, such as thickness, platform attributes, and dorsal scar patterns, however, generally remain unchanged and can be used as indicators of bifacial thinning. Therefore, in this analysis, I explore the relative use of flake blanks derived from bifacial thinning in the nonlocal tool assemblage as compared to the assemblages of debitage and local tools. It is expected that transported tools should be preferentially made on bifacial thinning flakes to maximize the transport efficiency of mobile toolkits.

Five measures are used to define the relative proportion of tools made on bifacial thinning flakes: interior platform angle, platform facetting, platform dulling, platform lipping, and a subjective categorical variable, "BFT," in which I made a judgment as to whether the tool blank was a bifacial thinning flake. The variable BFT was assigned a value of yes, no, or indeterminate.[1] Bifacial thinning flakes are expected to have greater interior platform angles and higher incidences of platform facetting, dulling, and lipping than "core reduction flakes," flakes produced from nonbifacial cores. Each variable is quantified for transported tools (those made on nonlocal raw materials) and compared to tools made on local raw materials and unmodified flakes larger than 2 cm in maximum dimension (table 6.1).

The results provide little support for the model. Most troubling, only 13 percent of tools made on nonlocal raw materials could be confidently determined to have been manufactured on bifacial thinning flakes. A chi-square test did find significant differences between the proportions of bifacial thinning and other flakes among tools (local and nonlocal) and debitage (table 6.1) but suggests that bifacial thinning blanks are underrepresented among both transported tools and those made locally. It appears that although bifacial thinning flakes were utilized as tool blanks in mobile toolkits, they were avoided for most tool forms, because there are greater proportions of bifacial thinning blanks within the debitage assemblage than the tool assemblage. Platform facetting and dulling are most common among nonlocal tools, but these differences were not

TABLE 6.1. Technological attributes of tools and debitage.

| Attribute | Nonlocal tools | Local tools | Flakes > 2 cm | Stat. test | Probability† |
|---|---|---|---|---|---|
| Platform facetted | 16 (.36) | 27 (.25) | 339 (.27) | | |
| Platform not facetted | 28 (.64) | 81 (.75) | 938 (.73) | $\chi^2$ | $\chi^2 = 2.27, p = 0.32$ |
| Platform dulled | 21 (.48) | 38 (.35) | 409 (.32) | | |
| Platform not dulled | 23 (.52) | 70 (.65) | 868 (.68) | $\chi^2$ | $\chi^2 = 5.07, p = 0.08$ |
| Platform lipped | 6 (.14) | 17 (.16) | 215 (.17) | | |
| Platform not lipped | 38 (.86) | 91 (.84) | 1,062 (.83) | $\chi^2$ | $\chi^2 = 0.38, p = 0.83$ |
| BFT = Y | 12 (.13) | 12 (.06) | 414 (.17) | | |
| BFT = N | 52 (.55) | 130 (.61) | 1,288 (.54) | $\chi^2$ | $\chi^2 = 19.33, p < 0.001$ |
| BFT = I | 30 (.32) | 69 (.32) | 699 (.29) | | |
| Avg. interior platform angle (deg.) | 124.2 | 116.3 | 119.8 | ANOVA | $F = 5.76, p = 0.003$ |
| Avg. thickness (mm) | 4.89 | 5.21 | 4.25 | | |

*Note*: †Chi-squared probabilities are one-tailed, and ANOVA probabilities are two-tailed.

significant at the .05 level (table 6.1). Platform lipping is fairly constant across all three categories of artifacts and shows no significant differences between the three categories (table 6.1). It is difficult to interpret artifact thickness between tools and debitage since variation in thickness is likely largely controlled by variation in mean artifact size, but it is interesting to note that the mean thickness for nonlocal tools (4.89 mm) exceeds that of the debitage sample (4.25 mm). Although this likely results from the nonlocal tool sample being composed of larger flake blanks, it may also suggest that little effort was made to select exceptionally thin flakes for blanks. The only support for the model is that significant differences were found among the groups for the mean interior platform angle, which was highest for nonlocal tools at 124.2 degrees (ANOVA, $F = 5.76$; $p = .003$). Aside from this one variable, there is little evidence that bifacial thinning flakes were favored for tools transported between sites.

The failure of the model to predict the predominant or preferential use of bifacial thinning flakes as blanks for tools transported between sites indicates that transport efficiency was not a primary design consideration. Of course, flake tools are transported to perform specific tasks, and therefore it may be unrealistic to expect portability to be a primary concern, even for mobile hunter-gatherers. Extremely thin flake tools obviously come with costs, most importantly that they are easily broken. Since tools in mobile toolkits are designed to be transported over large distances and for long periods of time, it is possible that thicker flakes were intentionally chosen to increase the durability and use-lives of transported tools. Furthermore, the added cost of carrying slightly thicker tools may have been minimal. If mobile toolkits carried by individuals consisted of hundreds or thousands of flake tools, then lightening the load by selecting very thin blanks might be expected, but if the population of tools and blanks transported numbered only in the tens, then the extra transport cost incurred by using more robust tools may have been slight. In contrast to portability, the design of flake tools for mobile toolkits may have been guided by currencies related to function and durability, but portability may have been an important constraint. It is important to note that this analysis does not falsify the Kuhn (1994) model. It only falsifies the hypothesis that transport efficiency was the primary currency governing flake tool design in this particular case.

## The Design of Toolkits: Cores versus Tools

In the previous section I explored the design of individual artifacts. In this section I look at the design of toolkits, or aggregates of artifacts, using a modification of the Kuhn (1994) model. Following Kuhn, I use the model to ask whether a forager should carry cores or flake tools or some combination of both when attempting to maximize utility per unit mass.

The transport cost of cores ($c_c$) is calculated in a similar manner to that of flake tools:

$$c_c = l_c^3 \tag{6.7}$$

where $l_c$ is the length of a core, and therefore cores are models as cubes. The utility of cores is assumed to be proportional to the volume of the core (because larger cores can produce more usable flakes) minus the size of a minimum usable portion, or slug:

$$u_c = r(l_c^3 - m_c^3) \tag{6.8}$$

where $m_c$ is the length of the core slug and $r$ is a coefficient of core efficiency. The term $r$, allowed to vary between zero and 1.0, describes the efficiency of reduction and incorporates both the amount of waste

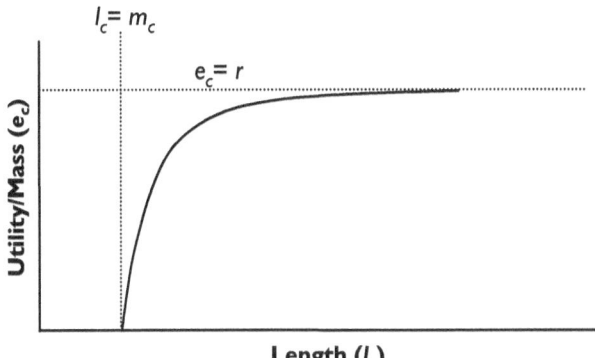

FIGURE 6.5. Graph of equation 6.9 showing transport efficiency vs. length for cores. Transport efficiency rises rapidly from the minimum usable size and approaches an asymptote at $e_c = r$.

produced through shaping and platform preparation and the number of blanks produced per unit volume. As $r$ approaches 1.0, reduction nears perfection, with virtually no stone going to waste, except that of the slug.

Taking the ratio of utility to transport cost provides a measure of core transport efficiency ($e_c$):

$$e_c = \frac{r(l_c^3 - m_c^3)}{l_c^3}$$

$$e_c = r\left(1 - \frac{m_c^3}{l_c^3}\right) \qquad (6.9)$$

In contrast to flake tools, transport efficiency increases continuously as a function of length for cores but approaches an asymptote at $e_c = r$ (fig. 6.5). Therefore, according to the model, transported cores should be as large as possible to maximize the usable portion of the core relative to the size of the slug. As cores become very large and approach their maximum efficiency, however, only slight improvements in transport efficiency are gained with increased size. Not surprisingly, the model also predicts that cores should be designed to minimize waste in reduction (larger values of $r$) and in the core slug (smaller values of $m_c$).

As the core model is formulated, it is difficult to compare the relative transport efficiency of cores and flake tools because they are expressed in terms of different units, specific to each artifact class. Here I modify Kuhn's core model to make it directly comparable to that of flake tools. The utility of a core ($u_c$) instead is modeled as the sum of the utility of its products, in this case a set of $n$ flake blanks:

$$u_c = n(l_t - m_t) \tag{6.10}$$

where $l_t$ is the length of those blanks, and $m_t$ is the length of the tool slugs. For simplicity, I use the linear model of flake tool utility and assume that $l_t$ and $m_t$ are constant, but none of these assumptions affect the predictions of the model. The transport cost of cores is calculated as:

$$c_c = nfl_t^3 + w_c \tag{6.11}$$

where $w_c$ is the amount of raw material wasted in the production of tool blanks. The term $w_c$ incorporates the terms $r_c$ and $m_c$ from Kuhn's original core model. In this framework, the transport efficiency of cores ($e_c$) is calculated as:

$$e_c = \frac{n(l_t - m_t)}{nfl_t^3 + w_c} \tag{6.12}$$

Using this model of a core, it is much simpler to compare the relative transport efficiencies of a set of flake tools.

To reframe the problem in a simple way, imagine you are gearing up for an extended period of time away from sources of lithic raw material, and you have a core that can produce twenty flake tools. Before you leave, should you produce the twenty tools or blanks and carry those, or should you transport the core and produce the flake tools from the core as you need them? Obviously there are many factors that could come into play when making that decision, but if the only consideration is transport efficiency, the solution is simple. To calculate the transport efficiency of a toolkit ($e_k$) composed of $n$ tools requires only the addition of a constant to the numerator and denominator of equation 6.4 (linear utility):

$$e_k = \frac{n(l_t - m)_t}{nfl_t^3} \tag{6.13}$$

where $n$ is equal to the number of tools to be carried. Notice that the number of tools carried has no impact on the toolkit since $n$ cancels out of the equation. Using equations 6.12 and 6.13, it is possible to ask under what conditions the transport efficiency of carrying a core is equal to or exceeds that of carrying the flake blanks ($e_c \geq e_k$):

$$\frac{n(l_t - m_t)}{nfl_t^3 + w_c} \geq \frac{n(l_t - m)_t}{nfl_t^3}$$

$$w_c \leq 0 \qquad (6.14)$$

From equation 6.14, it is clear that the transport efficiency of a core can equal that of its products if and only if reduction perfectly translates tool stone into usable products. If any waste is produced at all, then the more efficient solution is to carry the tool blanks. Since waste will be produced in core reduction in all circumstances, to maximize transport efficiency the most efficient solution will always be to carry finished tools and or blanks. Kuhn (1994:435) comes to a similar conclusion using a graphical portrayal of the original model. This prediction is so intuitively obvious that a mathematical demonstration of this fact is unnecessary, but in my mind it is one of the most important findings of the Kuhn model because it comes with a clear prediction of the archaeological record: cores should not be transported between sites if transport efficiency is the only factor influencing toolkit design.

## Testing the Toolkit Model

Any chipped stone artifact can serve as a core in the sense of serving as a medium for the production of flakes. This definition of "core" is similar to that of Andrefsky's (1998:10–11) "objective piece," but it is not the definition I use here. Instead, I follow Bamforth and Becker (2000:279), who define cores as "objects whose flaking patterns indicate reduction designed to produce useful flakes rather than to shape the worked piece into a useful form." This definition is necessary, by and large, to permit differential classification of bifaces and cores that take on a bifacial form. Although leaving the distinction to a subjective "judgment call" is certainly problematic, this is a judgment that is commonly made throughout the Americas in designating artifacts to types and is not unique to this study. I do not mean to imply that artifacts classified as bifaces were not used as cores, as they occasionally were, but for this analysis I am limiting the term "core" to those artifacts that I judge to have considerable utility as cores and relatively little tool utility. Bifaces are treated separately later in the chapter.

TABLE 6.2. Counts of cores by site and distance to source area.

| | Raw material source area | | | |
|---|---|---|---|---|
| Site | Local | Nonlocal | Unknown | Reference |
| Study sample | | | | |
| Carter/Kerr-McGee | 1 | 0 | 0 | |
| Agate Basin | 3† | 0 | 0 | |
| Krmpotich | 3 | 0 | 1 | |
| Barger Gulch, Loc. B | 31 | 0 | 0 | |
| Upper Twin Mtn | 1 | 0 | 0 | |
| Subtotal | 39 | 0 | 1 | |
| Expanded sample | | | | |
| Bobtail Wolf | 278 | 0 | 2 | Root et al. 2000:241, table 66 |
| Lake Theo | 2 | 0 | 0 | Buchanan 2002 |
| Big Black | 104 | 0 | 0 | William 2000 |
| Lindenmeier | 11 | 5‡ | 1 | Wilmsen and Roberts 1984:114, table 48 |
| Hell Gap, Loc 1. Folsom-Goshen level | 4 | 0 | 0 | Sellet 1999 |
| Elida | 0 | 15* | 1 | Hester 1962 |
| Stewart's Cattle Guard | 0 | 2 | 1 | Jodry 1999 |
| Blackwater Draw | 0 | 0 | 3 | Hester 1972 |
| Subtotal | 399 | 22 | 8 | |
| Total | 438 | 22 | 9 | |

*Notes*: †Frison (1982b:70) notes that five cores were recovered from the component, but only three were located for analysis. ‡The five cores from Lindenmeier classified as nonlocal may be derived from local jasper sources. *Although Hester (1962) classifies these artifacts as cores, they are more likely flake tools.

Although the data necessary to test the model were presented earlier in table 3.1, I repeat them in table 6.2 to spare the reader from an exercise in intensive page thumbing. The data provide strong support for the model. In this sample of cores, there is no evidence suggesting that cores were transported between sites. Of forty-two cores analyzed in the study sample, forty-one of those are manufactured on lithic raw materials local to the site area (within 20 km). The remaining core, which is from the Krmpotich site, is made on a lithic raw material of unknown origin. The most likely source is the Farson gravels underlying the Killpecker dune field and is therefore likely local to the site area.

This finding begs the question of whether there are any examples of cores manufactured on nonlocal lithic raw materials from Folsom or Goshen sites. An expanded sample was created using published data from nine additional Folsom sites (table 6.2). This sample includes 429 additional cores. Of those, 399 (93 percent) were derived from local raw material sources, eight are from unknown raw material sources, and 22 are from nonlocal sources. These data would, at first glance, suggest that cores were occasionally transported between sites, but much of the evidence supporting that idea is suspect. For example, Hester (1962) reports 15 cores made on nonlocal raw materials from the Elida site, but his description of the core assemblage makes one suspicious of whether these artifacts are cores at all: "Pieces of waste stone which may be considered as cores are rare and small in size. . . . There is no characteristic form. It appears that every possible flake was struck off before the core was discarded; these are not prepared cores of any specific type but appear to be the remnants of stone flaked somewhat at random. *In a normal lithic assemblage, these artifacts would be considered waste flakes.* In the present collection, however, there are no other candidates for cores" (Hester 1962:108, emphasis added). The five cores from Lindenmeier classified as "nonlocal" are manufactured on jasper, which Wilmsen and Roberts (1984) believe is predominantly derived from nonlocal sources, but they also note that "there is a very small jasper outcrop adjacent to a nearby chalcedony source, but, because of its color, it cannot account for more than a small proportion of the jasper inventory" (Wilmsen and Roberts 1984:114).

Unfortunately, it is not reported whether the jasper cores from Lindenmeier are derived from the local or more distant source areas. Because most of the jasper at Lindenmeier is nonlocal, I have assigned these cores to the nonlocal source, but they easily could have been classified as "unknown." The two remaining cores made on nonlocal raw material are from Stewart's Cattle Guard site, both of these being manufactured on Black Forest petrified wood (Jodry 1999:152). Jodry (1999:152) refers to one of these pieces as a "biface" but likely classified it as a core because of its robust width-to-thickness ratio of 2.2. The second artifact appears to be a true core and therefore is the only unquestionable example of a Folsom transported core. Although one could argue that this core alone could falsify the model, one confirmed nonlocal core out of a sample of 469 artifacts suggests transport efficiency did strongly shape toolkit

TABLE 6.3. Technological attributes of local and nonlocal debitage larger than 2 cm.

| Attribute | Nonlocal flakes | Local flakes | Stat. test | Probability† |
|---|---|---|---|---|
| Platform facetted | 51 (.52) | 313 (.24) | | |
| Platform not facetted | 47 (.48) | 983 (.76) | $\chi^2$ | $\chi^2 = 36.7, p < .001$ |
| Platform dulled | 56 (.57) | 376 (.29) | | |
| Platform not dulled | 42 (.43) | 920 (.71) | $\chi^2$ | $\chi^2 = 33.7, p < .001$ |
| Platform lipped | 22 (.22) | 206 (.16) | | |
| Platform not lipped | 76 (.78) | 1,090 (.84) | $\chi^2$ | $\chi^2 = 2.9, p = .091$ |
| BFT = Y | 124 (.56) | 300 (.12) | | |
| BFT = N | 39 (.18) | 1,406 (.58) | $\chi^2$ | $\chi^2 = 290.2, p \ll .001$ |
| BFT = I | 58 (.26) | 715 (.30) | | |
| Avg. interior platform angle (deg.) | 129.7 | 118.9 | Student's t | $t = -7.9, p < .001$ |
| Avg. thickness:width | 0.11 | 0.18 | Student's t | $t = 15.2, p < .001$ |
| Avg. thickness:length | 0.11 | 0.20 | Student's t | $t = 8.0, p < .001$ |

Note: †Chi-squared probabilities are one-tailed, and t-test probabilities are two-tailed.

design, at least with respect to the preference of transporting flake tools and blanks over cores.

Is it possible that cores were transported between sites, but were only rarely discarded in them? Certainly if the reduction of cores made on nonlocal raw materials occurred within sites, debitage from those cores should be present in sites. Are there core reduction flakes produced on nonlocal raw materials present in the sites in the study sample? Table 6.3 reports combined technological attributes of debitage larger than 2 cm from the study sample. In this sample, thirty-nine flakes manufactured on nonlocal raw materials were classified as core reduction flakes. This may suggest the presence of cores of nonlocal raw materials at the sites in the study sample with those cores remaining unrecovered or not discarded. Another possibility is that these flakes could have been misidentified. For example, these flakes could represent the early stages of bifacial reduction, in which the debitage produced takes on the character of core reduction. An analysis of reduction stage for flakes larger than 2 cm made on nonlocal raw materials provides some support for this idea (table 6.4). Debitage identified as core reduction flakes on nonlocal raw materials

TABLE 6.4. Reduction stage for nonlocal debitage larger than 2 cm.

| | Bifacial thinning flake? | | | |
|---|---|---|---|---|
| Attribute | BFT = N | BFT = Y | Stat. test | Probability[†] |
| >3 dorsal flake scars | 18 (.47) | 108 (.88) | | |
| ≤3 dorsal flake scars | 20 (.53) | 15 (.12) | $\chi^2$ | $\chi^2 = 27.9, p \ll .001$ |
| Pieces with cortex | 7 (.18) | 3 (.02) | | |
| Pieces lacking cortex | 31 (.82) | 120 (.98) | $\chi^2$ | $\chi^2 = 12.7, p < .001$ |
| Platform plain | 14 | 14 | | |
| Platform facetted or dihedral | 9 | 40 | Fisher's exact | $p = 0.004$ |

Note: [†]All probabilities are one-tailed.

have fewer dorsal flake scars, greater frequencies of cortex, and fewer dihedral and facetted platforms than debitage identified as bifacial thinning flakes. All of these differences are highly significant (table 6.4).

Thus the data provide strong support for one prediction of the Kuhn model. Tools and tool blanks were favored over the potential to produce tools (i.e., cores) in mobile toolkits. Carrying a core over hundreds of kilometers means carrying added weight in the form of wasted raw material, and prehistoric hunter-gatherers were certainly aware of this fundamental property of lithic reduction. It appears, therefore, that core reduction primarily occurred in close proximity to raw material sources, and tools and tool blanks were prepared at those locations for transport over long distances. Limited evidence exists that cores were transported between sites on rare occasions. The occasional inclusion of cores in mobile toolkits may be explained by the use of cores as heavy tools (Kuhn 1994:436–437), something not incorporated in the model. I return to this issue later, but the notion of an artifact that can be used as a core and a tool brings us to the question of the role of bifaces in mobile toolkits. Bifaces are often seen as the quintessential core tool, and in the next section I develop a model to explore the design of bifaces.

## The Design of Bifaces

A model of biface design must combine aspects of the core and tool models presented above. I first develop independent models of bifacial

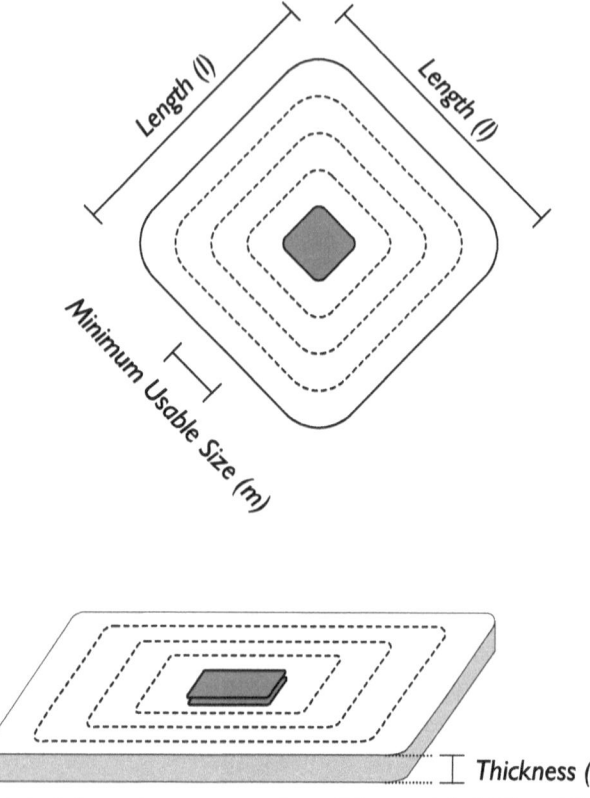

FIGURE 6.6. Schematic representation of the modeled biface. The shaded portion of the tool represents the unusable portion, and the unshaded portion represents the usable portion. Dashed lines represent edge positions after successive episodes of resharpening.

cores and tools and then combine these into a comprehensive "core tool" model. Bifaces are modeled as square in outline, but their thickness is permitted to vary independently (fig. 6.6). Although length and width could be modeled as independent variables, to do so would add unnecessary complexity to the model because it would not impact the predictions of the model whatsoever.

## Bifaces as Tools

If bifaces are designed solely to be used as tools, utility is calculated in a similar manner to Kuhn's (1994) flake tool model, but in contrast, it is assumed that the entire circumference of the biface is used as opposed to a single edge. Therefore, bifacial tool utility ($u_{bt}$) is calculated as:

$$u_{bt} = l^2 - m_t^2 \qquad (6.15)$$

where $l$ is the length of the biface and $m_t$ is the length of its minimum usable size. The transport cost for a bifacial tool ($c_{bt}$) is calculated as:

$$c_{bt} = l^2 t \qquad (6.16)$$

where $t$ is the thickness of the tool. The transport efficiency for a bifacial tool ($e_{bt}$) is calculated as:

$$e_{bt} = \frac{l^2 - m_t^2}{l^2 t}$$

$$e_{bt} = \frac{1}{t} - \frac{m_t^2}{l^2 t} \qquad (6.17)$$

The relationship between length and transport efficiency for bifacial tools is similar to that of cores in the Kuhn model where utility increases toward an asymptote at $e_{bt} = 1/t$. (fig. 6.7a). As would be expected, the model also predicts that bifacial tool utility increases as biface thickness and the minimum usable size decrease (fig. 6.7b, c). Therefore, if bifaces are designed to maximize transport efficiency and to be used only as tools, they should be made as large (in terms of length and width) and as thin as possible.

## Bifaces as Cores

The utility of a biface used as a core ($u_{bc}$) should be proportional to its volume of usable stone and therefore is modeled as:

$$u_{bc} = x(l^2 t - m_c^2 t_m) \qquad (6.18)$$

where $x$ is a coefficient of reduction efficiency, $l$ and $t$ are the length and thickness of the biface upon manufacture, and $m_c$ and $t_m$ are the minimum usable length and thickness of the core slug. The term $x$ is used to translate from a volumetric to a linear measure of utility and could, for example, be expressed in the units cm of edge produced per liter (or gram) of raw material reduced. Using this coefficient permits the direct comparison of the bifacial core and tool models.

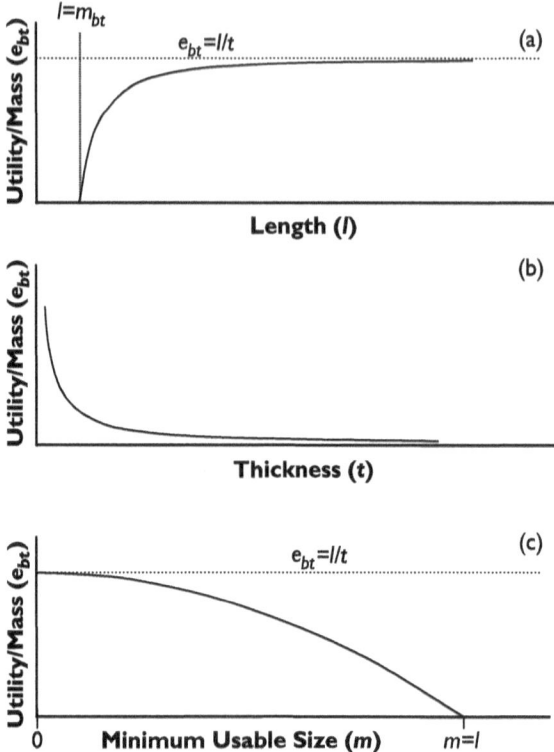

FIGURE 6.7. Graphical representation of equation 6.17 showing the modeled relationship between (a) length, (b) thickness, and (c) minimum usable size to transport efficiency for bifacial tools.

The transport cost for a bifacial core ($c_{bc}$) is calculated as:

$$c_{bc} = l^2 t \tag{6.19}$$

Therefore, the transport efficiency of a bifacial core ($e_{bc}$) is calculated as:

$$e_{bc} = \frac{x(l^2 t - m_c^2 t_m)}{l^2 t} \tag{6.20}$$

Like the bifacial tool model, the core model predicts that transport efficiency will increase continuously with greater length relative to the minimum usable size (fig. 6.8a). Unlike the bifacial tool model, when

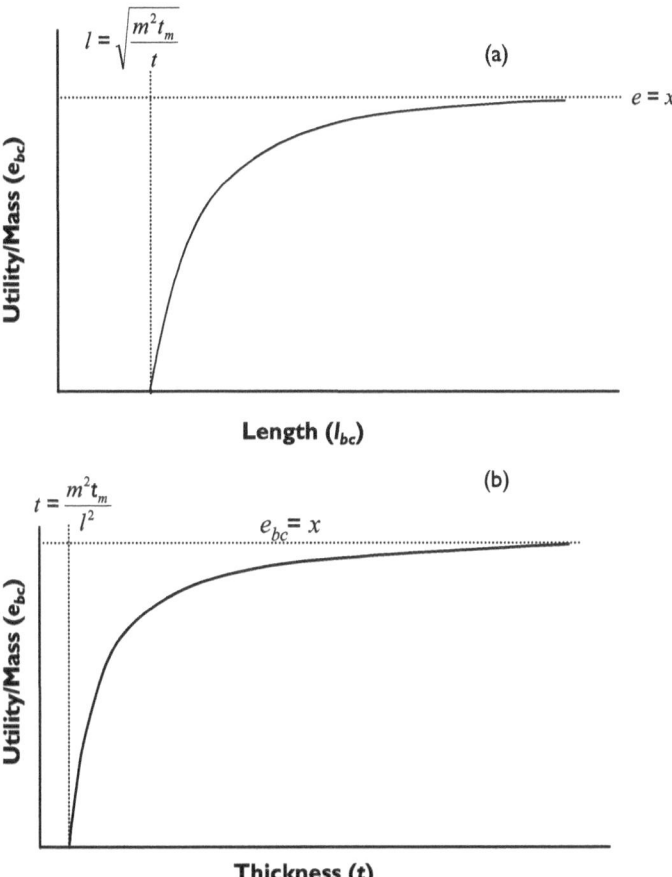

FIGURE 6.8. (a) Generalized graphical representation of equation 6.22 showing the modeled relationship between (a) length and (b) thickness and transport efficiency for bifacial cores.

using bifacial cores the model predicts that utility also increases with greater thickness (fig. 6.8b). Therefore, if bifaces are designed to be used solely as cores, they should be made as large as possible and, in contrast to bifacial tools, as thick as possible. The maximum thickness of a biface is constrained by the mechanics of bifacial reduction. When the edge angle of a bifacial core platform approaches 90 degrees, increases in thickness make further reduction impossible, thereby placing a physical constraint on maximum thickness.

## Bifaces as Core Tools

When bifacial tools and cores are modeled independently, there appears to be a conflict in design. A biface used as a tool optimizes transport efficiency when it is very thin to maximize the amount of edge relative to volume, and a biface used as a core should be made as thick as practical to maximize the amount of usable volume relative to the size of the slug. How should bifaces be designed if they are to be used as both tools and cores?

Combining the two models requires additional assumptions. Because an exhausted bifacial core still has tool utility (Morrow 1996; Rasic and Andrefsky 2001), the minimum usable length of a bifacial core should exceed that of a bifacial tool. It is assumed that the minimum usable length of a bifacial tool is smaller than that of a bifacial core ($m_c < m_t$), and that a biface smaller than the minimum usable core size (in terms of length and thickness) has only tool utility. Under the latter condition, the combined model will be identical to that of the bifacial tool model, and biface design should conform to the predictions of that model. If, however, a biface is larger than the minimum size of a bifacial core ($l \geq m_c$ and $t \geq t_m$), the utility of a bifacial core tool ($u_b$) is calculated as the sum of the utilities of bifacial cores ($u_{bc}$) and tools ($u_{bt}$):

$$u_b = l^2 - m_t^2 + 2x(l^2 - m_c^2)(t - t_m) \text{ when } l > m_c \text{ or } t \geq t_m \quad (6.21)$$

The transport cost of a bifacial core tool ($c_b$) is identical to that of the previous models, being a function of artifact size:

$$c_b = l^2 t \quad (6.22)$$

Therefore, the transport efficiency of a bifacial core tool ($e_b$), when it is larger than the minimum usable core size, is modeled as:

$$c_b = \frac{l^2 - m_t^2 + x(l^2 t - m_c^2 t_m)}{l^2 t}$$

$$e_b = x + \frac{1}{t} - \frac{m_t^2}{l^2 t} - \frac{x m_c^2 t_m}{l^2 t} \text{ when } l > m_c \text{ or } t \geq t_m \quad (6.23)$$

Although equation 6.23 is fairly complex, it retains aspects of both prior models. Because the length of a biface upon manufacture ($l$) appears only in the denominator of two terms that are subtracted from transport efficiency, larger bifacial core tools (in terms of length and width) have

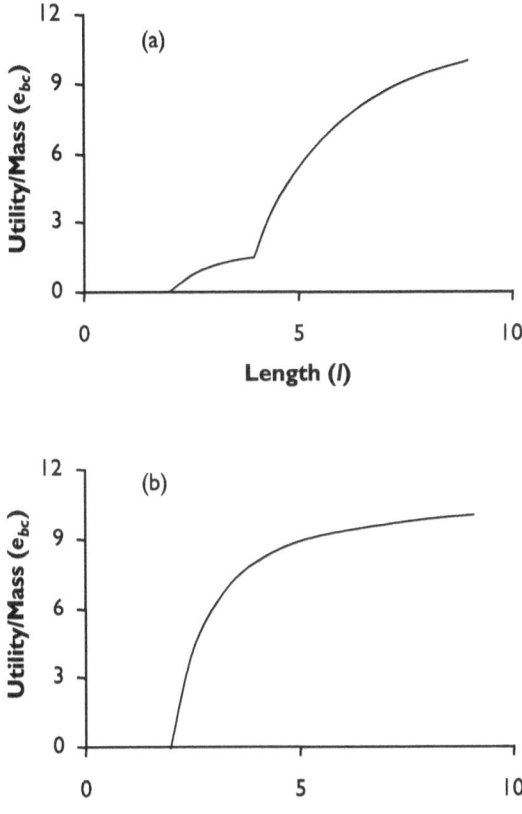

FIGURE 6.9. (a) Graphical representation of equation 6.23 showing the modeled relationship between length and transport efficiency for bifacial cores assuming $t = 0.5$, $x = 10$, $m_t = 2$, $m_c = 4$, and $t_m = 0.5$. The inflection point in the function occurs where the core tool has sufficient length to have core utility. (b) Same, except assuming $t = 2$. No step occurs because the biface retains core utility in its thickness since $t > t_m$.

greater transport efficiency (fig. 6.9). What about thickness? According to the model, under some conditions bifaces should be made as thin as possible, and under other conditions as thick as possible. To explain this prediction, we need to further evaluate equation 6.23.

Equation 6.23 describes the relationship between various aspects of biface design and transport efficiency. The slope of the function with respect to any of the independent variables describes whether an increase in that variable results in an increase or decrease in efficiency. Since the equation has a positive slope with respect to length, for example, increases in length result in increases in transport efficiency. The slope of the equation with respect to changes in thickness, however, is dependent on what values are plugged into the other variables. To show this, we take the first derivative of equation 6.23 with respect to thickness ($t$) because the derivative of the function gives the slope of the equation for any given value of thickness:

$$\frac{de_b}{dt} = \frac{m_t^2}{l^2 t^2} + \frac{xm_c^2 t_m}{l^2 t^2} - \frac{1}{t^2}$$

$$\frac{de_b}{dt} = \frac{1}{t^2}\left(\frac{m_t^2}{l^2} + \frac{xm_c^2 t_m}{l^2} - 1\right) \text{ when } l > m_c, \text{ or } t \geq t_m \quad (6.24)$$

Next, we ask under what conditions an increase in thickness results in an increase in transport efficiency, or under what conditions equation 6.23 has a positive slope with respect to thickness ($d_{eb}/d_t > 0$):

$$\frac{1}{t^2}\left(\frac{m_t^2}{l^2} + \frac{xm_c^2 t_m}{l^2} - 1\right) > 0$$

$$x > \frac{l^2 - m_t^2}{m_c^2 t_m} \text{ when } l > m_c \text{ or } t \geq t_m \quad (6.25)$$

Equation 6.25 describes the condition where increases in biface core tool thickness result in increases in biface core tool transport efficiency. According to this model, the decision to make a biface thick or thin will depend on the efficiency of bifacial reduction at producing usable tool blanks ($x$), the length of the biface upon manufacture ($l$), and the dimensions of the bifacial tool and core slugs ($m_t$, $m_c$, and $t_m$). Since the dimensions of the slugs can be considered constant, I focus my discussion on the other two variables, $x$ and $l$.

On the left side of the equation, if bifacial reduction is very inefficient at producing blanks (a small value of $x$), all things being equal, a biface should be made as thin as possible to maximize tool utility. In contrast, if reduction is very efficient (a large value of $x$), bifaces should be made as thick as possible to increase core utility. On the right side of the equation are the dimensions of the tool and core slugs and the length of the biface. The model predicts that if bifaces are small in terms of length and width (relative to the sizes of the slugs), then they should be made as thick as possible. If they are large relative to the dimensions of the tool and slug, they should be made as thin as possible. In this sense, we might expect there to be two optimal designs for bifacial core tools, either to make them small and thick or large and thin. Therefore, the thickness of a biface should hinge on the efficiency of bifacial reduction and a design decision, namely the size of the biface in terms of length and width.

This bifurcation in biface design is shown in figure 6.10. To understand these two design alternatives, we have to consider when an increase in thickness will result in an increase in transport efficiency. Imagine a biface that is 20 cm in length and width and 2 cm in thickness. Increasing the thickness of that biface to 4 cm will not impact its tool utility because its total edge potential remains constant. On the other hand, its core utility will increase, as will its transport cost, because both are a function of the amount of raw material in the biface. Only when an increase in core efficiency outweighs an increase in transport cost will greater thickness lead to an improvement in transport efficiency. This condition is only met when the length of the biface is small in comparison to the size of the tool and core slugs.

Although the model seems to predict a bifurcation in biface design, it actually predicts a trifurcation because equation 6.25 applies only when the dimensions of a biface exceed that of the core slug (assuming $t \leq t_m$). If the biface begins its use-life at a size smaller than the core slug (e.g., a biface manufactured on a small flake blank), it will take on the characteristics of the bifacial tool model and should be made as thin as possible. Figure 6.11 is a property space map of optimal thickness as predicted by the bifacial core tool model with respect to reduction efficiency ($x$) and length ($l$). With respect to thickness, the model predicts: (1) If bifacial core tool length is very large relative to the length of the core slug, bifaces should be made as thin as possible; (2) if length is small but larger than that of the core slug, bifaces should be made as thick as possible; and (3) if length is less than that of the core slug, bifaces should be as thin as possible. With respect to reduction efficiency, if bifacial core tools are extremely inefficient at producing usable flake blanks, bifaces should always be made as thin as possible. If reduction efficiency is high, even relatively large bifaces should be made as thick as possible.

These results may be somewhat counterintuitive. It is probably difficult to imagine small thick bifaces as optimally designed core tools. To address the relative advantages of small thick and large thin bifaces,

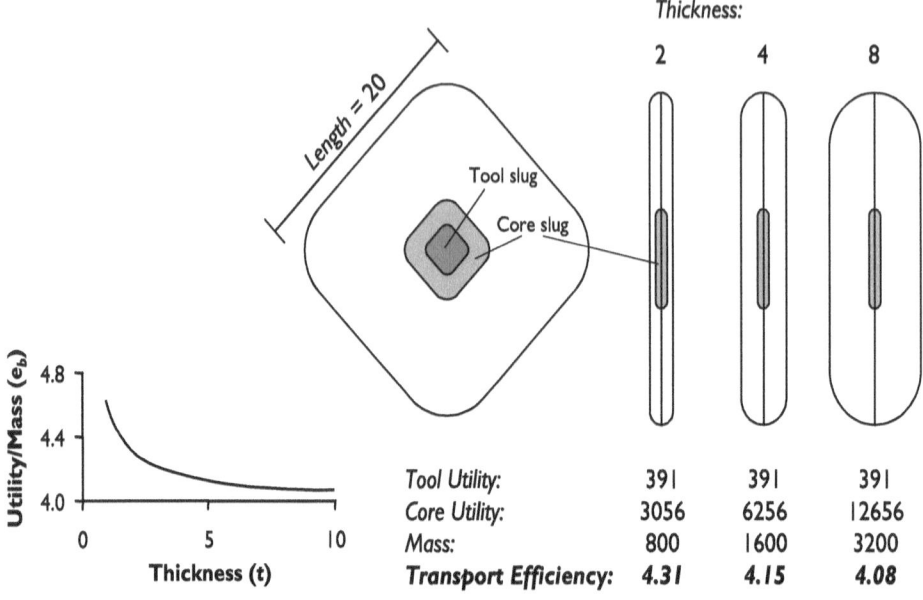

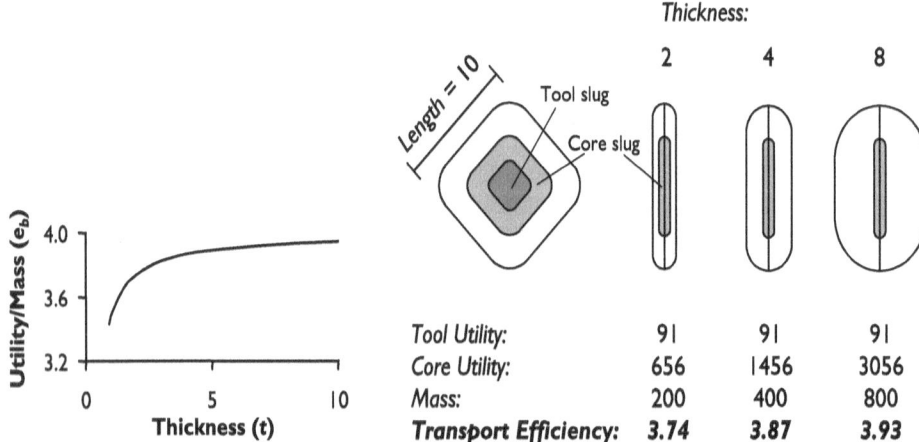

FIGURE 6.10. Graphical representation of the bifurcation in bifacial core tool design as predicted by equations 6.23 and 6.25. (top) A large biface loses transport efficiency ($e_b$) as thickness is added to the tool. (bottom) A small biface gains transport efficiency as thickness is added to the tool.

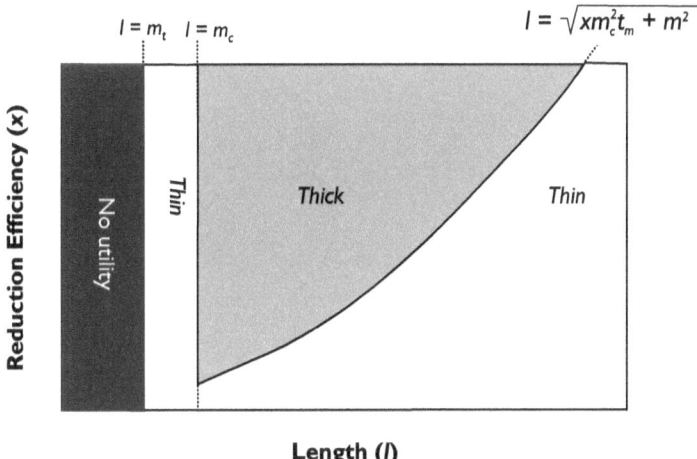

FIGURE 6.11. Property space map of the bifacial core tool model showing optimal thickness as a function of length ($l$) and reduction efficiency ($x$). In the area marked "thin," bifaces should be made as thin as possible to maximize transport efficiency. In the area marked "thick," bifaces should be made as thick as possible to maximize transport efficiency.

we can reframe the problem within the framework of the model. Instead of considering length and thickness independently, we can consider them in tandem and ask whether the transport efficiency of a bifacial core tool of a constant weight or volume, say 800 ml, is greater for a core tool that is small and thick or one that is large and thin. This example is explored in figure 6.12. If volume (or mass) is held constant, a large thin biface is always the optimal solution. With no change in volume, core utility and transport cost are not affected by the relative allocation of raw material to artifact length or thickness. Tool utility, on the other hand, is increased as the biface is lengthened, and therefore a large thin bifacial core tool will always have greater transport efficiency than a small thick bifacial core tool of equal volume.

## Summary

Although the models were designed to be simple, their predictions unfortunately are not. The most straightforward prediction of all three models (tool, core, and core tool) is that bifaces should always be designed to

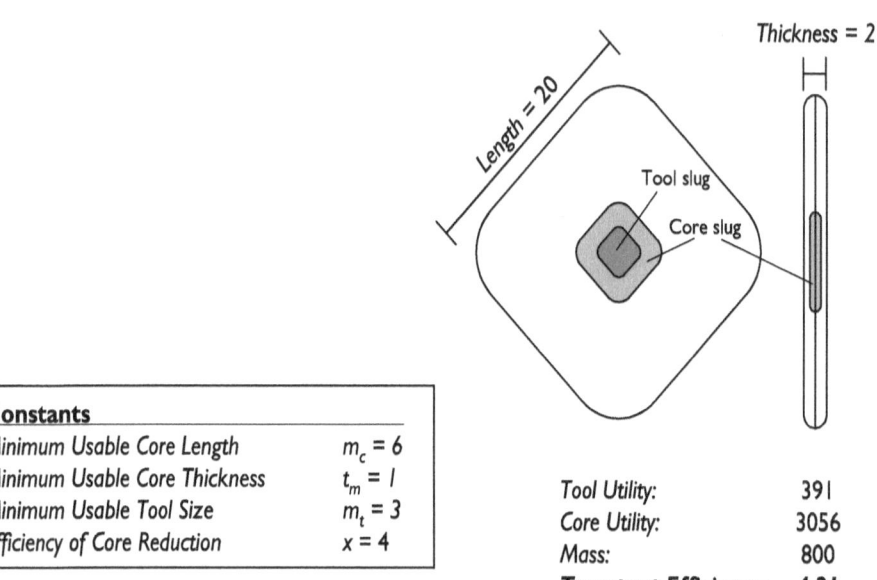

**Constants**
Minimum Usable Core Length    $m_c = 6$
Minimum Usable Core Thickness    $t_m = 1$
Minimum Usable Tool Size    $m_t = 3$
Efficiency of Core Reduction    $x = 4$

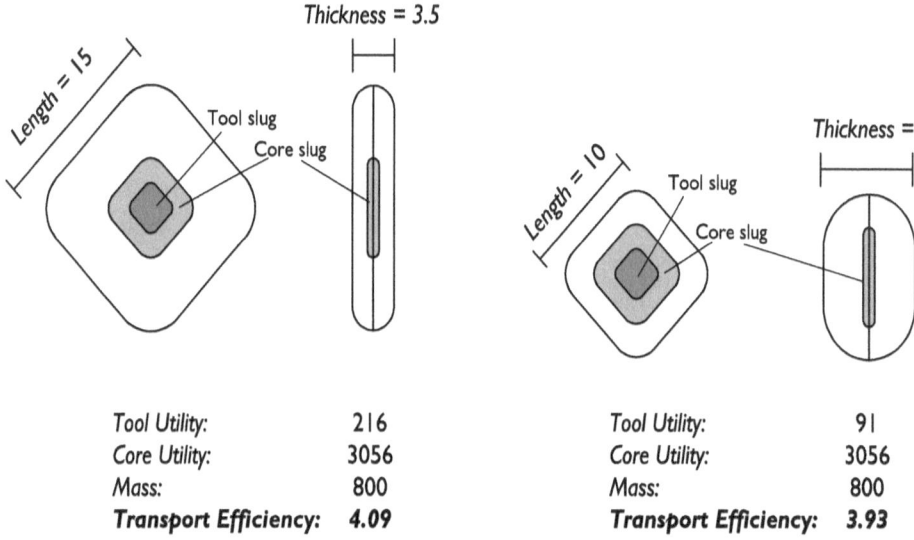

FIGURE 6.12. The transport efficiency of bifacial core tools holding volume (or mass) constant. Transport efficiency is optimal for large thin bifaces because greater allocation of stone to core tool length results in increased tool utility.

be as large as possible, with respect to length and width, if transport efficiency is the sole currency governing design. When area is added to a biface, in all cases it will have greater tool and core utility relative to transport cost. Of course, if bifaces get so large as to reduce their effectiveness as tools or hamper transport, their utility will decrease. Although such constraints are not built into the model, they would likely play an important role in limiting biface size.

The differences between the models relate to the variable thickness. A bifacial tool should always be made as thin as possible to maximize tool edge per unit mass. A bifacial core should always be made as thick as possible to maximize the amount of usable volume within that core. For a bifacial core tool, however, optimal thickness is dependent on: (1) the length (and width) of the biface upon manufacture, and (2) the efficiency of bifacial reduction at producing usable products.

The first step in producing a biface would be to determine the maximum length and width that a biface can take. Obviously, this will be constrained foremost by nodule or blank size and to a lesser extent by flintknapping skill and raw material quality. Nonetheless, the model predicts that the areal dimensions of a biface should always be maximized. The decision to make a biface thick or thin, again, depends on the efficiency of reduction and length as shown in figure 6.11. First, if the biface is so small that it has no core utility, it should be made as thin as possible because it will have only tool utility. If bifacial reduction is very inefficient at producing usable blanks, then bifacial core tools should always be made thin. If bifacial reduction is very efficient, then bifacial core tools should be made thin only if the biface is large relative to its minimum usable core and tool sizes. Therefore, if reduction is very efficient, and the biface is only slightly larger than its minimum usable size, it should be made as thick as possible. The model predicts two different outcomes depending on the starting conditions. Bifacial core tools should be large and thin, or small and thick.

Figure 6.12 demonstrates, however, that a large thin bifacial core tool always has greater transport efficiency than a small thick core tool of equal mass. In this respect, why would you ever manufacture small thick bifaces? According to the model, you would do so only if you were unable to manufacture large thin bifaces of equal mass. One obvious example would be a landscape in which large pieces of raw material are not available. If this constraint was lifted and large nodules of raw material were available, then the optimal bifacial core tool would always be large and thin.

*Paleoindian Bifaces in Context*

Testing the biface models developed above is extremely difficult because the models predict only how bifaces should be manufactured to maximize transport efficiency. Rarely do we recover stone tools in "pristine" form. Folsom bifaces are believed to have been serially reduced and sometimes recycled over extended periods of time (Ahler and Geib 2000; Boldurian and Hubinsky 1994; Hofman 1992; Jodry 1998; Root et al. 1999; Surovell, Waguespack, and Kornfeld 2003). Although these behaviors conform to the assumptions of the model, they also mean that bifaces recovered from archaeological sites are usually many steps removed from the initial manufacture event or broken during that event and therefore do not reflect the initial design parameters. Also, bifaces, unlike flake tools, usually do not preserve any characteristics of their state upon manufacture.

Numerous types of bifaces are recognized in Folsom assemblages (see chap. 2), including projectile points, preforms, ultrathins, and cores. Bifaces that would not fall into any of these categories are also found; these artifacts are similar in length and width to ultrathins but are somewhat thicker and may be bifacial tools that were intended to be transformed into projectile points (Nami 1999). For the Southern Plains, Hofman (1992) argues that all of these forms were integrated into a single technological system, centered on the use of large bifacial cores. In this model, Folsom stone workers geared up with large bifacial cores at sources of Edwards chert in west Texas and Alibates silicified dolomite in the Texas panhandle. These artifacts served as cores early in their use-lives, producing large flake blanks for the production of projectile points and flake tools. As they were further reduced, they would function as tools and would eventually be made into projectile points. Hofman sees reduction proceeding in steps occurring after a bison kill event. Extremely large bifaces are certainly known from the Plains (Stanford and Broilo 1981; Wyckoff 1996) but have never been recovered from a primary context within a Folsom site. One well-known artifact, commonly referred to "Frank's biface" (Hofman 1994:197), was found on the surface in the vicinity of the Mitchell Locality of Blackwater Draw (Stanford and Broilo 1981), but its association with the Folsom occupation has been questioned (Bamforth 2002b:68). Large bifaces have also been recovered from Clovis caches (Frison and Bradley 1999; Gramly 1993; Wilke et al. 1991). The largest of these is from the Anzick cache, measuring 31.6 × 17.0 × 1.7 cm (Wilke et al. 1991). Frank's biface is similar in size, but it is considerably thinner relative to its width at 28.8 × 17.2 × 0.37 cm (Stanford and Broilo 1981). These artifacts are commonly referred to as "cores," presumably because they are believed to be too large to have effectively functioned as tools. If they were designed solely to be cores, however, their design would not conform to the predictions of the bifacial core model, wherein bifaces should be made as thick as possible. With width:thickness ratios of 10.0 and 46.5, respectively, they conform much better to the bifacial tool and core tool models. The extremes in length and width certainly are consistent with all of the biface models, and because they likely far exceed their minimum usable size, efforts to make them extremely thin compare well to the predictions of the core tool model.

Bifaces were occasionally used as cores (Hofman et al. 1990; Jodry 1999; LeTourneau 2001; Tunnell 1977), but bifacial thinning flake blanks are consistently rare in Folsom tool assemblages relative to core-struck flake blanks (table 6.1). This would seem to contradict the notion that bifacial reduction is very efficient in terms of converting stone into usable flake edges (Kelly 1988). In fact, Prasciunas (2004) found that bifacial reduction was no more efficient at producing usable flake edges than amorphous core reduction, and other studies have confirmed that Paleoindians seemed to have often eschewed bifacial thinning flakes for tool manufacture (Bamforth 2002b; LeTourneau 2001). With respect to the core tool model, a low value of $x$ (core efficiency) would favor the manufacture of extremely thin bifaces, and in this respect the large thin bifaces common in early Paleoindian assemblages may conform to the model very well.

While extremely large and thin bifaces do conform well to the predictions of the model, they may not conform perfectly to the assumptions of the model. For example, it is assumed that a bifacial edge is used as a tool from manufacture to discard. Oversized bifaces may not have been used as tools in their early stages simply because they were too large to be effectively manipulated for performing basic tasks. In this sense, it might be interesting to incorporate functional variables into the model beyond "minimum usable size." An experimentally derived function comparing size and work efficiency (see Morrow 1996) or a variable "maximum usable size" could certainly be added to the model. If these artifacts were designed to serve as cores early in their life cycles and tools later on, optimal biface design could certainly differ from the models developed above.

In addition to functioning as cores in the typical sense, large bifaces may have served as blanks for smaller biface (projectile point and ultrathin) manufacture through intentional segmentation by radial or snap fracture. Evidence from the Anzick site (Wilke et al. 1991), Barger Gulch (Surovell, Waguespack, and Kornfeld 2003), and the Lake Ilo sites (Root et al. 1999) provide strong support for this idea. A large biface, then, could be broken into a number of smaller biface fragments to be converted into other bifacial tool forms. In this sense, for early Paleoindians, large bifaces may have been very efficient cores for the production of usable bifaces but not particularly efficient in the production of usable flake tools.

Smaller biface forms (i.e., length $\leq 20$ cm) are common to Folsom assemblages, namely ultrathins, preforms, and projectile points. Like the "giant bifaces" discussed above, ultrathin bifaces generally have a width: thickness ratio exceeding 10:1 (Root et al. 1999). Whether large bifacial "cores" and smaller ultrathin bifaces served different technological roles is unknown, but the extreme thinness of ultrathins conforms very well to the bifacial tool model. Likewise, the design of Folsom projectile points could be interpreted as optimization of transport efficiency. Although they are not particularly large with respect to length and width, among Paleoindian projectile points Folsom points are easily the thinnest (Tankersley 1994), thereby maximizing usable edge to mass. Crabtree (1966:3) perceived what he called "classic" Folsom as "being as thin and perfectly shaped as the technique would allow." In this light, fluting could be seen only as a means of maximizing transport efficiency. The hypothesis that fluting was performed to reduce projectile point weight is almost as old as the Folsom discovery itself (Roberts 1936:19). I find this explanation of fluting somewhat unsatisfactory because it is difficult to believe that the gain in transport efficiency resulting from fluting would outweigh the risk of preform failure during manufacture.

In sum, Folsom biface technology may have been designed to maximize transport efficiency, but this conclusion is by no means straightforward. Early Paleoindian bifaces show extremes in length, width, and thickness that are predicted by the models. Large thin bifaces, commonly called cores, such as Frank's biface recovered from Blackwater Draw, are consistent with the core tool model, which predicts that bifaces should be made as long and wide as possible to maximize tool and core utility. Bifaces did not serve as the primary cores for the production of flake tools, although bifacial thinning flakes were occasionally used as blanks. The primary core function of large bifaces may have been in the production of smaller bifaces through intentional fracture and segmentation. Extremely thin projectile points and ultrathins are consistent with the bifacial tool model and the idea that these artifacts served primarily as tools (Jodry 1998, 1999; Root et al. 1999). If Folsom hunter-gatherers were extremely successful at optimizing transport efficiency through the control of bifacial thinning to produce large and thin bifaces, why do these strategies not persist into later time periods? As stated in chapter 2, extremes in mobility should result in extreme constraints on technology, and in this sense the scale of Folsom residential mobility may have placed a severe constraint on portability that was lessened in later periods when mobility was reduced. It is interesting to note, however, that bifaces may have been made ultrathin to "lighten the load," but flake tools were not. This contradiction may relate to the way these tools are used. Perhaps bifaces were most often used in a direction parallel to the tool edge, while most flake tools were used perpendicular to the tool edge. In the former case, a thickness would have less of an effect on the probability of breakage, and this idea is consistent with the use of projectile points and hypotheses that ultrathins functioned as knives (Jodry 1998, 1999; Root et al. 1999).

## *Bifaces, Flake Tools, and Cores in the Folsom/Goshen Toolkit*

In many ways and in many forms, Folsom bifaces seem to conform to the predictions of the models. The models assume a goal of maximizing portability or transport efficiency in biface design. Therefore, it is not surprising that bifaces were among the set of artifacts commonly carried between sites. Interestingly, bifaces morphologically unrelated to projectile point production (i.e., those that are not classified as preforms or points) are relatively rare on nonlocal raw material, being represented by only two of thirty-one specimens. In the study sample, bifaces made on nonlocal raw materials occur most commonly as preforms and projectile points. Ingbar (1992) notes similar patterning at the Hanson site and argues that it likely results from staging of biface manufacture. Bifaces that are transported into sites are more likely to exist at later stages of reduction than those made locally. The relative scarcity of nonlocal bifaces that have not approached the stage of point production probably is not a result of such artifacts not being transported between sites, but rather results from the fact that they rarely are discarded in the early to middle stages of reduction.

Paleoindian mobile toolkits were then composed of bifaces and flake tools. Flake tools were most commonly manufactured from blanks produced from blocky or amorphous cores, but those cores were not transported between sites, or only very rarely transported. The differentiation between local flake tool production and nonlocal biface reduction is reflected in the combined nonlocal debitage and tool assemblages from the study sample (table 6.3). Debitage produced on nonlocal raw materials is dominated by bifacial thinning debris, while debitage produced on local raw materials is dominated by core reduction. Statistically significant variation in this regard is evident in the frequencies of platform facetting and dulling, average interior platform angle, and the average ratios of thickness: width and thickness:length.[2] This pattern is also mirrored by the Hanson site assemblage (Ingbar 1992). The contradiction that flake tools transported into sites were infrequently produced on bifacial thinning flakes (table 6.1) and that the majority of debitage produced from artifacts transported into sites is produced by bifacial thinning is easily explained by the hypothesis that flake tool production, or at least flake tool blank production, largely took place within the vicinity of raw material source areas.

This idea contrasts with arguments by Bamforth and Becker (2000), who suggest that Paleoindians regularly transported cores between sites. They base this hypothesis on two lines of evidence. They argue that refits of debitage at the Allen site in Nebraska indicate that more cores were present at the site during its occupation than were recovered. The ratio of cores to bifaces was 0.05, but the ratio of refitted flake sequences produced on cores to that of bifaces was 1.17. Secondly, they note that numerous sites have produced core-struck flakes but no cores, attesting to presence but not recovery of "phantom" cores (Bamforth and Becker 2000:286). An alternative explanation of these patterns could be that early stage bifacial reduction flakes are identified as having been produced by core reduction. They argue further that the absence or low frequencies of cores in many Paleoindian sites can be explained by their long use-lives. This explanation, however, receives no support from core raw material data from a suite of thirteen Folsom and Goshen sites (table 6.2). As discussed in chapter 3, long-lived artifacts have only a very small probability of discard in sites with short occupation spans. If cores have long use-lives and are transported, in archaeological assemblages they should rarely be made on local raw materials except in very long occupations. Furthermore, it would be expected that long-lived cores on nonlocal raw materials would be likely discarded in sites with very long occupation spans. Neither of these expectations holds. For example, Barger Gulch, Locality B, and Bobtail Wolf, both sites believed to represent relatively long-term occupations, have a combined 311 cores, none of which are made on nonlocal raw materials (table 6.2). Thus, in contrast to Bamforth and Becker (2000), I suggest that cores were not an integral component of Paleoindian mobile toolkits, but they were an integral component of Paleoindian "immobile toolkits." They were used in settings where raw material was readily available but rarely moved beyond those locations. Carrying of cores translates to carrying of unnecessary raw material, which likely explains why they were left behind when camp was moved. This property of cores, however, leaves one thing unexplained. Why did cores ever leave the quarry? Why not produce flake tools or blanks at the quarry and maximize the transport efficiency of raw material brought into campsites?

There are many possible explanations for this behavior. The accumulation of a surplus of lithic raw material, as discussed in chapter 4, may be one explanation, but one could certainly accumulate a surplus of flake blanks, rather than cores. Another explanation is that cores provide more flexibility in flake tool production in contrast to producing a set of flake blanks at a quarry locality, which potentially limits the sizes and forms of available flake blanks brought to a residential location. Also, as mentioned above, cores can be used as heavy tools in chopping or similar tasks, which may occur at campsites (Kuhn 1994:436–437), but artifacts interpreted as heavy-duty tools in Paleoindian sites are commonly made of low-quality or non-cryptocrystalline varieties of stone (e.g., Bradley and Frison 1996:61; Ferring 2001:171; Frison 1982b:61–65; Frison and Bradley 1980:76–79; Root et al. 2000:238–239; Wilmsen and Roberts 1984:122), indicating that high-quality tool stone may have often been spared from those tasks. A better explanation of the transport of cores to campsites may relate to foraging or procurement efficiency.

Metcalfe and Barlow's (1992) model of field processing, introduced in chapter 1, suggests that distance plays an important role in the decision of whether to remove the unusable portion of a resource in the field or in camp. All things being equal, when lithic raw materials are close to a camp the model predicts that it is more efficient, in terms of utility gained per unit time, to transport unmodified nodules of lithic raw material to camp (see fig. 1.2). In other words, when you live close to a quarry area, the time saved in transporting unmodified nodules outweighs the potential gains in transport efficiency that could be made by removing excess raw material at the quarry. The Metcalfe and Barlow (1992) and Kuhn (1994) models together provide an elegant explanation of why cores might be carried from quarry to camp, but not from camp to camp.

## Summary

In this chapter, using the Kuhn (1994) model of mobile toolkits and additional models derived from that work, I explored various aspects of tool, core, biface, and toolkit design assuming a goal of maximization of transport efficiency. Flake tools did not conform to the predictions of the model, suggesting that functional currencies may have been more important than transport efficiency in guiding flake tool design. Numerous aspects of early Paleoindian biface design, however, may be explained by efforts to optimize transport efficiency. Paleoindian mobile toolkits, the set of artifacts transported between residential sites, consisted exclusively of bifaces and flake tools. Flake tools were preferentially produced on flakes struck from cores, but in concert with the Kuhn (1994) model, core reduction was almost entirely restricted to campsites in the vicinity of lithic raw material sources. Because cores do not appear to have been regular constituents of mobile toolkits, debitage produced on nonlocal raw material is dominated by bifacial thinning debris. In contrast, nonlocal flake tools are rarely made on bifacial thinning flakes, suggesting that bifaces functioned primarily as tools.

# 7

# On the Optimal Production of Trash

THIS CHAPTER IS ABOUT lithic debitage, the cigarette butts and candy wrappers of the Stone Age. By debitage, I refer to unwanted, discarded, and abandoned by-products of the manufacture of wanted things, namely tools. The category "debitage" includes angular debris and complete and fragmentary flakes and pot lids. In numerous early excavations, debitage was abandoned by the prehistoric occupants of sites and by archaeologists who recovered it. During the excavations at Lindenmeier, for example, the great majority of lithic debitage was not collected. Boldurian and Cotter (1999), citing a 1988 correspondence to John Cotter from E. Lohr, report: "In 1940, toward the end of the final season at Lindenmeier, Roberts instructed one of his workers to bury into the backfill 'a huge sack of chips and flakes' that the dig had produced" (Boldurian and Cotter 1999:37).

Frank Roberts's treatment of debitage at Lindenmeier was by no means unique for this era. For example, Haury was similarly indifferent to debitage in his 1941–42 excavations of Ventana Cave (Huckell and Haynes 2003:356). Although debitage certainly has received short shrift in many studies of Paleoindian technology, it is no longer viewed with such apathy. Recent books and papers dedicated to the analysis of debitage show that this prehistorically unwanted debris now is recognized as an important tool for studying human behavior (Amick and Mauldin 1989; Andrefsky 2001; Hall and Larson 2004; Shott 1994).

This chapter differs in an important way from prior chapters. In chapters 5 and 6 I explored optimization of technological behavior with regard to lithic procurement, the accumulation of stone to be devoted to tool production, and tool design, decisions made in tool manufacture to optimize tool morphology with respect to some currency. This chapter differs in that I focus on the optimal discard of debitage, artifacts never intended for use. This idea of the optimization of garbage production may sound like a strange concept. If unused and unwanted debris presumably is simply discarded at the location where it is produced, presumably with little forethought, it is a behavior somewhat difficult to construe as a decision-making process. Although the discard of debitage itself may be difficult to construe as an optimality problem, it can also be considered the by-product of decision-making processes. Debitage is produced when objective pieces (e.g., cores, tools, bifaces, projectile points, etc.) are reduced. The reduction of any artifact presumably occurs for a reason, and one could certainly construe reduction as driven by optimization concerns (e.g., chap. 6 of this study, Brantingham and Kuhn 2001; Kuhn 1994; Metcalfe and Barlow 1992). One could ask, for example, when should a nodule of stone be reduced? Where should it be reduced? Should it be reduced to a biface, blade core, flake blanks, or something else? Should it be completely reduced now, or only when the need arises? Debitage is not only a direct reflection of such decisions, but also because it rarely moves far beyond the location where it was produced, unlike tools that were commonly carried from place to place, debitage is an abundant location-specific record of flintknapping behaviors. To model the "optimization of trash production," then, is to model the optimization of behaviors that produce trash. Specifically, in this chapter I attempt to explain intersite variability in frequencies of bifacial thinning and core reduction debris. I begin with a simple accumulation model, which is used to guide later analyses of debitage production. I argue that differences in debitage frequencies among the sites of the study sample can be explained in part by decisions governing the manufacture and use of flake tools.

## A Null Model of Debitage Frequencies

I build on models discussed in chapter 3 to model the discard of core reduction and biface thinning debitage. I refer to this model as a "null model" because it predicts that the proportions of different types of debitage should not respond to variation in occupation span. To show why this is, a useful point of departure is Schiffer's (1975a:840, 1987:53) discard equation:

$$d_t = \frac{S}{L} t \qquad (7.1)$$

where $d_t$ is the number of artifacts discarded as a function of time ($t$), $S$ is the number of a given type of artifact in systemic context, and $L$ is the average use-life. Applying this equation to chipped stone debitage is somewhat problematic, although I have done so in previous chapters. Most people would agree that the use-life of debitage is very brief (e.g., Bamforth and Becker 2000:284; Schlanger 1990), lasting only from the time the core is struck to the time the flake hits the ground since its "use" is only in its removal, its presence being an impediment to the morphology of a desired object. With respect to the discard equation, therefore, a very small value of $L$ would equate to a very large number of flakes being produced and deposited in archaeological sites, an observation that is certainly in accordance with the archaeological record. The term $S$, however, is more problematic. How many flakes, on average, are maintained systemically in the toolkit? While some usable tool blanks may be maintained, if not the answer to this question could easily be zero, and if so the answer could be very few. In this respect, no matter how short the use-life of a flake is, few flakes would be expected to be discarded, an observation not reflected by the archaeological record. These arguments are dependent on the assumption that flake use-life begins at the moment of flake production, but instead one could "start the timer" at the moment of raw material acquisition. In this case, we could estimate how many flakes, bound up in various forms of raw material, are maintained in the toolkit (the term $S$), but estimating average use-life would be problematic. Is the average use-life of a flake bound up in a core equal to one-half the use-life of that core? If so, then what is governing the use lives of cores? Cores are reduced and waste flakes are produced in response to other behavioral needs, not because the cores or waste flakes themselves are necessarily needed or used. Modeling debitage discard instead as a function of tool stone needs, as a by-product of tool, biface, and core production and reduction, is conceptually much more straightforward.

The discard of core reduction and bifacial thinning flakes must be modeled independently because they are produced for different reasons. Large bifacial thinning flakes will be produced during the production and reduction (maintenance/resharpening) of bifaces. Although cores could be modeled similarly, the great majority of the cores in the study sample are informal. For informal cores, manufacture and reduction can be considered synonymous and therefore do not need to be modeled independently. Also, by definition bifaces are reduced to produce bifaces, while cores are reduced to produce tools.

Beginning with bifaces, the number of thinning flakes produced in bifacial reduction over the course of an occupation ($n_{br}$) can be calculated as:

$$n_{br} = \frac{S_b t}{r} f_{br} \qquad (7.2)$$

where $S_b$ is the number of bifaces in systemic context, $t$ is the occupation span, $r$ is the frequency of bifacial reduction (the time elapsed between reduction episodes, or the average use-life of a bifacial edge), and $f_{br}$ is the number of bifacial thinning flakes produced per reduction episode. To model the discard of bifacial thinning flakes resulting from biface production, it is assumed that a new biface is manufactured when a biface in the toolkit is expended, lost, or broken. Therefore, the equation to model the discard of bifacial thinning flakes in biface production takes on the form of Schiffer's discard equation with the addition of a constant representing the number of flakes produced per production episode:

$$n_{br} = \frac{S_b}{L_b} f_{bp} \qquad (7.3)$$

where $S_b$ is the number of bifaces in systemic context, $t$ is the occupation span, $L_b$ is the average use-life of a biface, and $f_{bp}$ is the number of bifacial thinning flakes produced per manufacture episode. The total number of bifacial thinning flakes produced ($n_b$) over the course of an occupation of length $t$, then, is the sum of equations 7.2 and 7.3:

$$Nb = \frac{S_b t}{r} f_{br} + \frac{S_b}{L_b} f_{bp}$$

$$n_b = S_b t \left( \frac{f_{br}}{r} + \frac{f_{bp}}{L_b} \right) \qquad (7.4)$$

Turning now to cores, it is further assumed that core reduction flakes are produced solely in flake tool manufacture. Therefore, core reduction occurs in response to loss, breakage, or expending of flake tools in the toolkit. Therefore, the number of core reduction flakes discarded ($n_c$) for an occupation of length $t$ is calculated as:

$$n_c = \frac{S_t t}{L_t} f_t \qquad (7.5)$$

where $St$ is the number of flake tools in systemic context, $L_t$ is the average use-life of a flake tool, and $f_t$ is the number of flakes produced per flake tool manufacture episode.

To model the relative proportions of bifacial thinning and core reduction flakes discarded as a function of time, I take the ratio of equations 7.4 and 7.5:

$$\frac{n_c}{n_b} = \frac{\frac{S_t t}{L_t} f_t}{S_b t \left( \frac{f_{br}}{r} + \frac{f_{bp}}{L_b} \right)}$$

$$\frac{n_c}{n_b} = \frac{\frac{S_t}{L_t} f_t}{S_b \left( \frac{f_{br}}{r} + \frac{f_{bp}}{L_b} \right)} \tag{7.6}$$

For the time being, the most important property of equation 7.6 is that the variable occupation span ($t$) is cancelled out. According to the null model, therefore, the relative proportions of bifacial and core reduction flakes in an assemblage should remain constant with respect to occupation span. Because the rates of discard of both flake types are assumed to be constant with respect to occupation span, as shown in chapter 3 (equation 3.8), the relative proportions of bifacial thinning and core reduction debitage will remain unchanged.

## Testing the Model

The null model predicts that the ratio of core reduction to bifacial thinning flakes should be constant across a range of occupation spans. The data set of the study sample, then, is ideal for testing the model. As shown in chapter 3, the four Folsom campsites included in the sample span exhibit a wide range of mean per capita occupation spans. Following the analysis performed in chapter 6, for each site four measures are used to estimate the relative proportions of bifacial thinning and core reduction flakes. First, I quantify the relative proportions of flakes subjectively identified as bifacial thinning and core reduction flakes (the variable BFT introduced in chap. 6). These subjective categorizations are compared to quantitative

TABLE 7.1. Technological attributes of debitage by site for all artifacts analyzed independently.

| Attribute | Site | | | | |
|---|---|---|---|---|---|
| | CKM | AGB | KRMP | BGB | UTM |
| BF thinning flakes* | 40 | 276 | 237 | 338 | 1 |
| Core reduction flakes | 13 | 177 | 390 | 1,532 | 6 |
| Indeterminate flakes | 20 | 110 | 321 | 1,364 | 1 |
| CRF:BFT | 0.33 | 0.64 | 1.65 | 4.53 | 6 |
| Avg. interior platform angle (deg.) | 128.3 | 125.0 | 122.2 | 117.9 | 116.4 |
| Avg. length:thickness† | 9.72 | 9.13 | 7.68 | 6.71 | 5.73 |
| Avg. width:thickness‡ | 8.40 | 8.35 | 7.68 | 6.75 | 8.23 |

*Notes:* *Includes channel flakes. †Sample is limited to flakes where a reliable length measurement could be made. ‡Sample is limited to flakes where a reliable width measurement could be made.

TABLE 7.2. Correlation matrices of debitage attributes for two samples.

| | CRF:BTF | IPA | L:T | W:T |
|---|---|---|---|---|
| All sites | | | | |
| IPA | −.957 (.011) | | | |
| L:T | −.963 (.008) | .986 (.002) | | |
| W:T | −.415 (.487) | .538 (.350) | .468 (.427) | |
| Upper Twin Mountain Excluded | | | | |
| IPA | −.945 (.055) | | | |
| L:T | −.929 (.071) | .983 (.017) | | |
| W:T | −.988 (.012) | .957 (.043) | .966 (.034) | |

*Note:* Cells contain Pearson's correlation coefficients, with probability values in parentheses.

measures of mean platform angle, length:thickness, and width:thickness. A site dominated by bifacial thinning debitage should have greater average interior platform angles and ratios of width:thickness and length:thickness than a site dominated by core reduction debris. The sample of debitage includes all flakes that were analyzed individually (see Surovell 2003a: appendix 1). Table 7.1 presents a summary of the data, and table 7.2 shows Pearson's product moment correlation matrices of the four variables.

For the five sites, all of the indicators of the relative proportions of bifacial thinning and core reduction debitage are highly correlated except for average width:thickness. When Upper Twin Mountain (representing a sample of only six artifacts) is removed, however, width:thickness is significantly correlated with all variables. All variables are strongly correlated ($r > 0.9$ for all correlations), but two of the correlations are not significant. The directions of the correlations are consistent with expectations in that greater proportions of flakes identified as having been produced by bifacial thinning correlate negatively with average interior platform angle, length:thickness, and width:thickness. In general, this analysis suggests that with the possible exception of Upper Twin Mountain, debitage proportions are reflected by these four variables.

To compare occupation span and debitage frequencies, I combined the four debitage attribute variables to create a core reduction index representing the ratio of core reduction to bifacial thinning flakes for each site. Each variable was standardized to scale between 1 and 10 by linear interpolation between minimum and maximum observed values. For example, for the variable CRF:BFT (table 7.1), a value of 10 was set to the maximum observed value (6.00 for Upper Twin Mountain), and a value of 1 was assigned to the minimum observed value (0.33 for Carter/Kerr-McGee). The remaining values were scaled accordingly. To account for negative correlations, the reverse was done for the remaining three variables. The core reduction index is the average of the standardized debitage attribute values. Low values of the core reduction index imply a site dominated by bifacial thinning flakes, and high values imply a site dominated by core reduction. The standardized attribute values, core reduction index, and various measures of occupation span are presented in table 7.3.

The data provide no support for the null model. While the model predicts no correlation between occupation span and the core reduction index, they are highly correlated for the four campsites (fig. 7.1). A familiar pattern emerges, with Upper Twin Mountain standing as a lone outlier breaking the trend, a fact likely attributable to site function and small sample size. Longer occupations are increasingly dominated by debitage produced by core reduction. The core reduction index increases as a log function of the occupation span index (OSI) and artifact density. Both relationships are highly statistically significant (Ln[OSI] vs. core reduction

TABLE 7.3. Standardized debitage attribute values, the core reduction index, and measures of occupation span by site.

| | Site | | | | |
|---|---|---|---|---|---|
| Variable | CKM | AGB | KRMP | BGB | UTM |
| Standardized attribute values | | | | | |
| CRF:BFT | 1.00 | 1.49 | 3.10 | 7.67 | 10.00 |
| Avg. interior platform angle | 1.00 | 3.50 | 5.61 | 8.87 | 10.00 |
| Avg. length:thickness | 1.00 | 2.33 | 5.60 | 7.79 | 10.00 |
| Avg. width:thickness | 1.00 | 1.27 | 4.93 | 10.00 | 1.93 |
| Core reduction index | 1.00 | 2.15 | 4.81 | 8.58 | 7.98 |
| Measures of occupation span | | | | | |
| Occupation span index (OSI) | 0.26 | 1.13 | 8.25 | 100.00 | 1.45 |
| Ln (OSI) | −1.35 | 0.12 | 2.11 | 4.61 | 0.37 |
| Artifact density (per m$^2$) | 1.18 | 3.11 | 7.89 | 33.10 | 0.47 |
| Ln (artifact density) | 0.17 | 1.13 | 2.07 | 3.5 | −0.76 |

index, $r = 0.994$, $p = 0.006$; Ln[artifact density] vs. core reduction index, $r = 0.990$, $p = .010$). Thus, virtually 100 percent of the variability in debitage proportions among the four campsites can be explained solely by the duration of occupation of those sites.

*Discussion*

Although the null model was falsified, when asking where the null model went wrong there are obvious variables to consider. Recall from chapter 3 and above that if the rates of discard of two types of artifacts remain constant through the course of an occupation, then the proportion of those artifacts within an assemblage will remain constant. In this case, because the ratio of two artifact types changes as a function of occupation span, the rate of discard of at least one of those artifact types must change with respect to occupation span. Because core reduction flakes become increasingly dominant as occupations are prolonged, there are at least three possible explanations: (1) The discard rate of core reduction flakes increases with time; (2) the discard rate of bifacial thinning flakes decreases with time; or (3) some combination of the two. These alternative hypotheses are depicted graphically in figure 7.2. Unfortunately, a

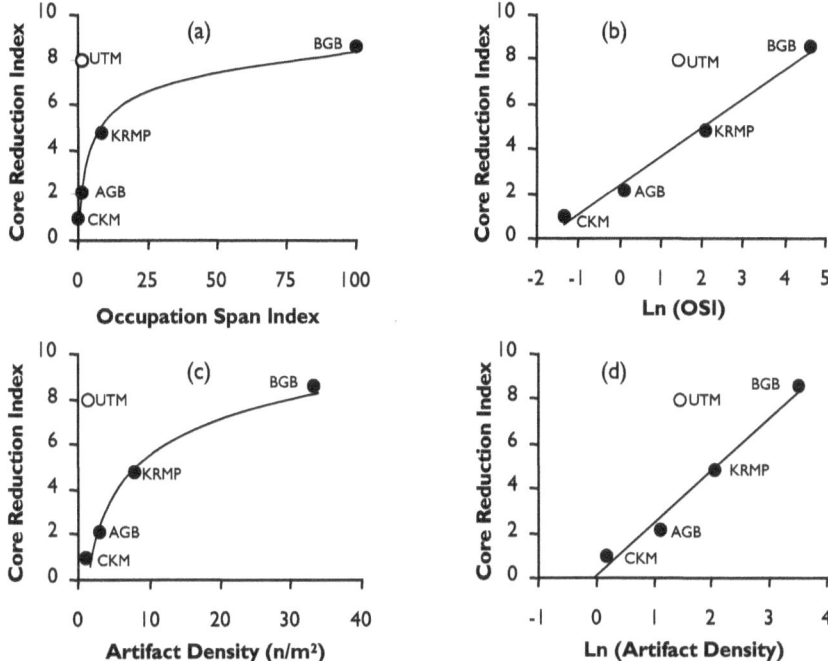

FIGURE 7.1. (a) Regressions of the occupation span index (OSI), (b) Ln (OSI), (c) artifact density, and (d) Ln (artifact density) vs. the core reduction index with Upper Twin Mountain (shown as open circle) excluded. Artifact density is the number of artifacts >2 cm per m².

direct test to distinguish between these three hypotheses is not possible. Any such test would require examination of subtle differences in curvilinearity (see fig. 7.2) that can be rigorously detected only with large sample sizes. With only four sites, performing such an analysis would inevitably produce ambiguous results. Nonetheless, indirect tests are possible because the null model suggests which variables should control the discard rates of core reduction and bifacial thinning flakes, and changes in any of those variables over the course of an occupation will result in changing rates of debitage discard.

According to the null model (equation 7.6), changes in any or all of eight variables could cause the pattern observed in figure 7.1. With respect to the discard of bifacial thinning flakes, the relevant variables are (1) the number of bifacial thinning flakes produced per biface manufacture event, (2) the number of bifaces maintained in the toolkit,

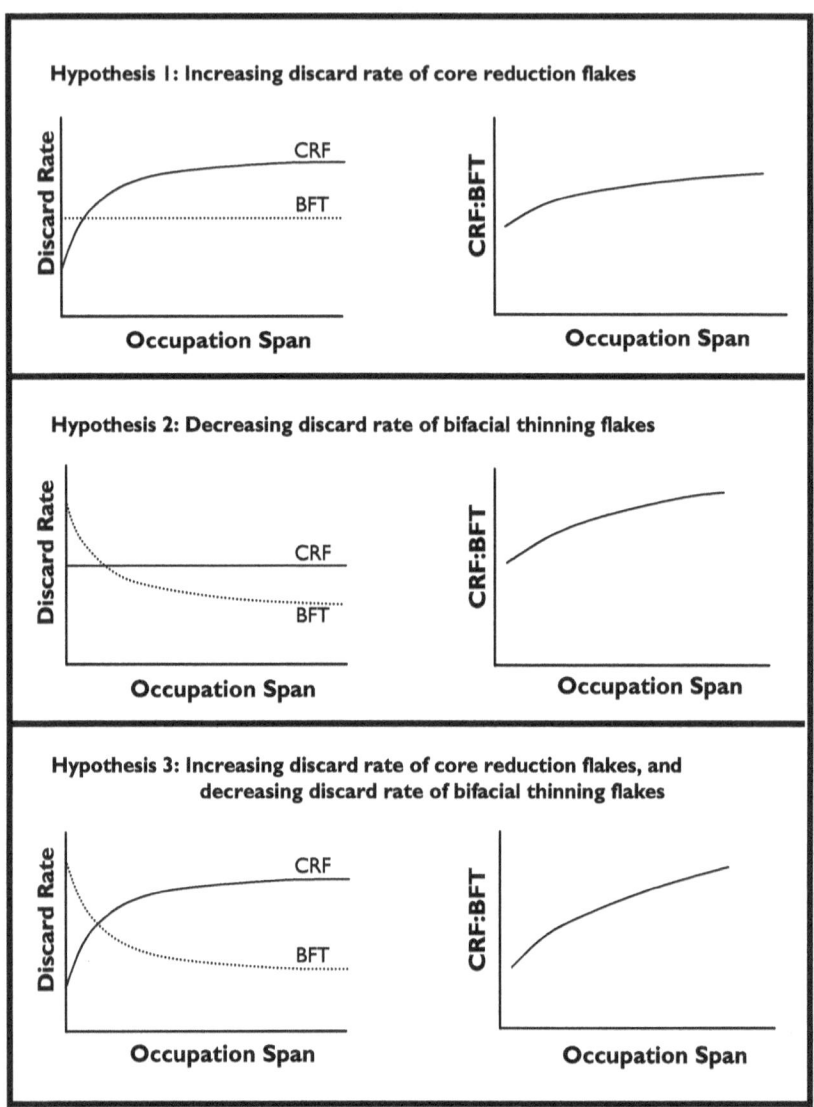

FIGURE 7.2. Three hypotheses to explain changing ratios of core reduction to bifacial thinning debris. The left graph of each pair shows the proposed relationship between occupation span and discard rate. The right graph shows the predicted relationship between occupation span and the ratio of core reduction to bifacial thinning flakes. All three hypotheses produce very similar archaeological signatures, resulting in a problem of equifinality in attempting to distinguish between them.

(3) the frequency of bifacial reduction for edge maintenance, (4) the number of bifacial thinning flakes produced per biface maintenance event, and (5) the average use-life of bifaces. With respect to the number of core reduction flakes produced, the relevant variables are (1) the number of flake tools maintained in the toolkit, (2) the average use-life of a flake tool, and (3) the number of core reduction flakes produced per flake tool manufacture event. Because all of these variables are independent, the number of potential combinations of variables affecting the system is daunting. Also, I can think of no a priori reason why any of these variables should necessarily respond alone to how long a person or a group of people stays in a single location—at least a location where lithic raw material is available to replace worn-out tools. For example, it would be very difficult to establish a link between the number of flakes produced per tool or biface manufacture event and occupation span without invoking some other factor. One "other factor," I argue, is raw material availability. Occupation span, per se, is not driving the system, but instead changes in raw material availability that correlate with occupation span.

## *What Is Raw Material Availability?*

The notion of raw material availability as a primary constraint shaping technological variability has received considerable attention (e.g., Andrefsky 1994; Bamforth 1986, 1990a,b; Holdaway et al. 1996; Kuhn 1994, 1995; Marks et al. 1991; Nelson 1991). Most often, raw material availability is conceived of as a natural property of environments or locations. In other words, some locations have more available lithic raw material than others, independent of human activity. People, therefore, adjust their technological strategies to cope with a dearth or abundance of raw material encountered on a landscape. Many people have noted, however, that raw material availability is also a product of human behavior—not simply an external environmental factor of which humans are aware and to which they react. For example, Nelson writes: "If raw material is unavailable, it is because humans have made social, economic, and technological decisions that create the condition. For example, they have settled at great distance from stone sources and have not stockpiled stone" (Nelson 1991:77).

Others have noted a direct relationship between the frequency of residential mobility and the availability of lithic raw materials because long stays in a single location permit the stockpiling of tool stone:

> The duration of individual occupations can itself be expected to affect the relative cost of obtaining raw materials. The longer people remain at a location, the more trips made out from it, the greater the opportunities to procure raw materials from the surrounding countryside. (Kuhn 1991:78–79)

> An important difference between mobile hunter-gatherers who make expedient tools where raw materials are abundant and sedentary peoples lies in the availability of lithic raw material. . . . Even where the necessary raw material is not locally available, however, sedentary populations may continue to make expedient tools, and undertake regular trips to quarries so that raw materials can be stockpiled. (Parry and Kelly 1987:301)

> More sedentary people can stockpile stone for future use, removing the need to spend their time and energy in extensive production of sophisticated tools. (Bamforth 1990a:70)

Generally speaking, it is argued that frequent residential mobility over large distances equates to an unpredictable supply of tool stone, and therefore low raw material availability, because you have only that which you can carry on your back and you may not know the exact itinerary of your trip. Residential stability, on the other hand, allows stockpiling of lithic raw materials and therefore permits raw material abundance even in contexts of natural raw material scarcity. Raw material availability, then, is proportional to both the amount of stone on hand and the relative cost of acquiring new stone from natural sources. In this sense, raw material availability could be high for a person hundreds of kilometers away from high-quality sources of stone, a condition that holds for many flintknappers today who load pickup beds full of stone and stockpile raw material at their residences. Likewise, a person only a few miles away from a source of raw material could feel raw material stress if their immediate raw material supplies were dwindling. From a site-specific perspective, raw material availability, then, is both a natural (nonhuman) and cultural property. While the geologic source areas cannot be moved, nodules of raw materials can be, and as occupations are prolonged, "cultural source areas" can be created by lithic stockpiling, potentially affecting many aspects of technology.

Although stockpiling of lithic raw material is certainly made possible by residential stability, this does not explain why it occurs. In chapter 5, I have already presented evidence for the accumulation of surplus lithic raw materials for the sites of the study sample. I argued that stockpiling occurred to cushion against lithic shortfalls that require extra procurement trips for the acquisition of additional lithic raw materials. Although this is certainly not the only possible explanation (e.g., Kuhn 1995; Parry and Kelly 1987), I reiterate it here to note that the stockpiling of tool stone is potentially costly and therefore would be favored only in contexts where some advantage is gained by doing so.

Can raw material availability be construed, then, as simply a function of natural and cultural raw material abundance? Unfortunately it is not that simple because the rate at which stone is consumed will also impact availability. To give a simple example, imagine you have a jug with two liters of water, and the closest natural source of water is 200 km away. This would seem like a fairly dire situation for a pedestrian, but it is dire only if water is being consumed at a rapid rate. If you require only 10 ml of water per day, for example, you have enough water to last 200 days. If you require one liter per day, you have enough for only two days. From this simple example, it should also be clear that the rate at which you are moving across the landscape should affect availability as well. Consider the difference in availability for a person who is walking versus one who is driving in a vehicle. In the latter case, the availability would be higher because the natural source will be reached sooner, and we would expect the behavior of individuals in each situation to reflect those differences. A person who is immobile is perhaps in the most dire of situations but again has the unique capability of amassing stores. Ranchers and farmers far from gas stations, for example, commonly keep stores of fuel in tanks at their residences to reduce procurement costs. More familiarly, archaeologists far from towns stockpile food, water, and supplies at their camps for the same reason.

Although a simple formal model of raw material availability can be constructed using these principles, I leave that exercise for a later date. The point is that raw material availability is considerably more complex than

TABLE 7.4. Core density and core mass density for the sites of the study sample.

| | Site | | | | |
|---|---|---|---|---|---|
| Variable | CKM | AGB | KRMP | BGB | UTM |
| Cores (n) | 1 | 5† | 4 | 31 | 1 |
| Total core mass (g) | 46.3 | 509‡ | 389.25 | 4,768.11 | 79.18 |
| Excavation area (m²) | 34 | 171 | 74 | 51 | 25.5 |
| Core density (no./m²) | 0.03 | 0.03 | 0.05 | 0.61 | 0.04 |
| Core mass density (g/m²) | 1.36 | 2.98 | 5.26 | 93.49 | 3.11 |

*Notes:* †Only three cores were located during the analysis of the Agate Basin assemblage. This number is based on Frison's (1982b:70) count. ‡The mass for cores from Agate Basin was calculated using the observed mass for the three cores (305.53 g) analyzed and multiplying by 5/3 to account for the masses of the two missing cores.

only the amount of stone available within a catchment area surrounding a site. It should respond to the natural and cultural quantities of stone available, the rate of consumption of stone, and, for mobile peoples, the rate at which you are approaching a natural source of raw material. What, then, is raw material availability? I would define the phrase as the inverse of the cost, measured in terms of time or energy, of replacing a toolkit or components thereof. It must be the inverse of replacement cost because high raw material availability implies that it is relatively inexpensive (low cost) to replace a toolkit. For the purposes of this chapter, I focus primarily on one component of raw material availability, the amount of stone stockpiled at a site, which should be negatively correlated with replacement cost.

If raw material availability is in part a function of the amount of stone stockpiled at a campsite, then one critical variable would be the quantity of stone bound up in the form of cores because cores served as the primary source for flake tools. Here, I again test the proposition that the longer one stays in a given location, the greater the number and mass of cores that will be stockpiled. Figure 7.3 shows the relationship between the OSI and the density of cores and raw material bound up in cores (table 7.4). Excluding Upper Twin Mountain, a Pearson's correlation of the natural logarithms of the occupation span index and core mass density is significant at the .05 level ($r = 0.960$, $p = 0.040$), but a comparable

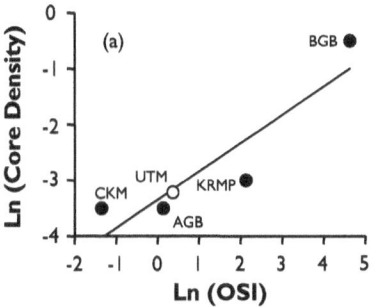 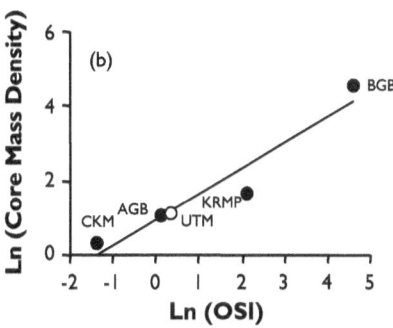

FIGURE 7.3. (a) Scatterplot of the Ln (occupation span index) vs. Ln (core density [the number of cores per m²]) for the five sites of the study sample. (b) Scatterplot of the Ln (occupation span index) vs. Ln (core mass density [the total mass {g} of cores per m²] for the five sites of the study sample. Regressions exclude Upper Twin Mountain (shown as open circle).

correlation with the log of core density is not ($r = 0.907$, $p = 0.093$). Because raw material mass is a better measure of availability than numbers of cores, I do not consider this problematic. Therefore, this relationship would suggest that raw material does become increasingly available as occupations are prolonged.

## *When Does a Flake Tool Live to Old Age?*

In the previous chapter I demonstrated that the toolkit entering sites is composed exclusively of bifaces and flake tools/blanks. In this respect, perhaps it should not be surprising that large debitage in short-term occupations is dominated by flakes produced from bifaces, because at the initiation of an occupation the greatest potential to produce large debitage is in the form of bifaces. This fact by itself, however, does not necessarily explain changing debitage frequencies, because upon arrival at a campsite a toolkit can be quickly supplemented with cores made from local raw material and the discard of core reduction flakes could then keep up with the discard of bifacial thinning flakes. In fact, not only were cores accumulated quickly in campsites, but also the mass of stone acquired consistently exceeded the amount needed, thereby creating surpluses.

What effects could this buildup of raw material mass have on technological behavior? Here I focus on flake tool use-life. From equation 7.6, if the average use-life of flake tools (the variable $L_t$) decreases as occupations are prolonged, then tools and core reduction debitage will be produced at an accelerating pace. This could produce the observed pattern of changing debitage frequencies (fig. 7.1). Would the average use-life of flake tools be expected to decrease as raw material becomes increasingly available? Numerous individuals have suggested as much (e.g., Holdaway et al. 1996; Kuhn 1995:33–34; Rolland and Dibble 1990). Here I formalize this idea by turning to a classic model from foraging theory—the marginal value theorem (Charnov 1976b; Stephens and Krebs 1986:24–32).

The marginal value theorem is used in ecology and anthropology to explain the length of time that a forager should remain in a patch given the distance between patches and the relationship between the time spent in a patch and energetic returns. Figure 7.4a is a graphical depiction of the model. The relationship between returns and time spent in a patch is deemed the gain function. It is generally assumed that the rate of energy capture declines as a function of time spent within a patch because the patch becomes depleted, so the gain function is typically curvilinear with decreasing slope. The optimal solution for time spent in a patch is solved by determining where a line, drawn intersecting the x-axis at the travel time between patches, is tangent to the gain function. The model predicts that as the time (or distance) between patches is increased, a forager should remain longer in a patch. Although the marginal value theorem was developed as a description of foraging behavior, there is no reason why it cannot be applied to the discard of stone tools. Figure 7.4b shows that a simple substitution of variables allows its application. The typical gain function for a patch of food, or any exhaustible resource for that matter, mimics that of a stone tool. As a stone tool is progressively used, its edges dull, and therefore it will become less effective the longer it is used. As it is resharpened it also loses mass and edge length, further reducing its effectiveness. Therefore, the functional efficacy of a stone tool should decline as its use-life is prolonged (see Beck et al. 2002; Bousman 1993:fig. 14; Prasciunas 2004:figs. 8 and 9). Therefore, in figure 7.4b, I have simply substituted tool use-life for time spent in patch, work performed for total return, and the time required to replace a tool for travel time. The variable work performed is the total amount of work performed by a tool and could be expressed, for example, in the units $cm^2$ of hide scraped, cm of meat cut, or $cm^3$ of wood carved.

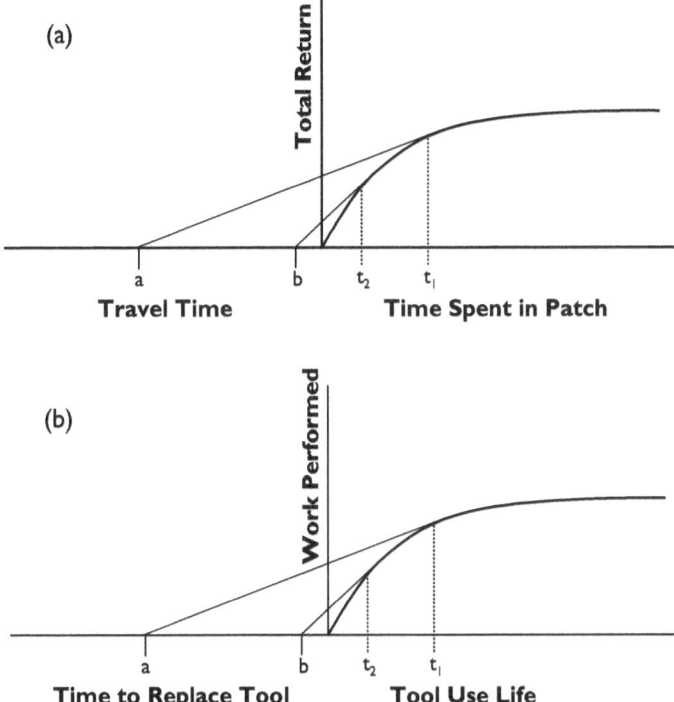

FIGURE 7.4. (a) The marginal theorem patch choice model. On the right side of the x-axis is the time spent in a patch, and on the left side of the x-axis is the travel time between patches. The y-axis is the amount of the resource acquired. The heavy line is the gain function describing the relationship between the time spent in a patch vs. the amount of resource acquired. The optimal patch time is determined by the tangential intersection of a line drawn through the travel time. The figure shows two optimal solutions ($t_1$ and $t_2$) for different travel times (a and b). The model predicts that as travel time increases, so should the time spent foraging within a patch. (b) The marginal value theorem as applied to flake stone tools. The model takes the same form, with the variable tool use-life substituted for time spent in patch, work performed for total return, and the time required to replace a tool for travel time. Likewise, the model predicts that as more time is required to replace tools, tools should be discarded later in their use-lives.

The time required to replace a tool would take into account not only the time required to manufacture and possibly haft a new tool, but also the time to acquire the raw material to do so.

Although raw material might be immediately available in the toolkit, I argue that this variable should represent the time-averaged cost (measured in time) of replacing a tool, taking into account the future raw material costs imposed by discarding the current tool. For example, if you are 100 km from a source of stone and have only two extra tool blanks on hand to replace a tool currently in use, the immediate time to replace a tool would be very short, perhaps a few seconds. However, when the third and final tool is discarded, the time required to replace that tool could be on the order of days because a 200-km round-trip would be required. Therefore, we can expect tool discard to reflect not only immediate replacement costs but future replacement costs as well. In this respect, the variable time required to replace a tool, when time-averaged, is essentially a measure of raw material availability.

The marginal value theorem applied to stone tools, therefore, predicts that, all things being equal, the average use-life of tools should decrease as the time to replace tools decreases, or as raw material availability increases. The exact form of the relationship will depend on the form of the gain function, and I dare not predict the exact shape of the gain function for the average Paleoindian tool (nor that of use-life and occupation span). However, cutting experiments by Prasciunas (2004: figs. 8 and 9) suggest that a gain function like that shown in figure 7.4 is likely appropriate for relating artifact size to use efficiency (e.g., cm cut/minute). It is predicted then that as occupations are prolonged and raw material availability increases, flake tools will be discarded earlier in their use-lives.

## Testing the Model

To test the application of the marginal value theorem to flake tool discard, it is necessary to derive measures of flake tool use-life. It is assumed that the degree of retouch exhibited on tools is proportional to use-life. That is, tools with greater retouch intensity were discarded at later stages in their use-lives. These measures are intended to be proportional to the length of time from manufacture to discard, rather than the residual use-life, or the amount of potential use time remaining in the tool upon discard. Three measures are used. The first is the variable retouch intensity, subjectively assigned a value of 1, 2, 3, or i. A value of 1 was given to tools with very slight retouch on only a portion of the tool edge, a value of 3 was given to tools with very invasive retouch covering all or most of the tool edge, and a value of 2 assigned to intermediate cases. A tool was assigned a value of i (indeterminate) if only a small portion of tool (such as an edge fragment) was recovered. The second measure uses the ratio of the thickness of the retouched edge relative to the thickness of the tool. Edge thickness was measured in three locations on each tool and averaged (Surovell 2003a:340–53). A lightly retouched tool should have a relatively thin working edge relative to the thickness of the tool itself. Finally, for each tool, the amount of edge exhibiting retouch was quantified by dividing tool edge into ten radial edge units (see Surovell 2003a:appendix 1). The presence or absence of retouch on each portion was noted, as was the presence of a fresh or break edge. Break edges were subtracted from available edge units since the absence of retouch on a break edge does not speak to the amount of retouch on the tool prior to breakage. For each site, I quantified the average retouch intensity, average edge:tool thickness, and percentage of the available edge units utilized (tables 7.5 and 7.6). All three variables are highly correlated for the five sites of the study sample (for all correlations: $r > 0.9$; $p < 0.05$). Figure 7.5 shows the distribution of retouch by edge unit for the five sites. The four campsites show a similar pattern. Retouch is least common adjacent to the platform and gradually increases in frequency to the distal portion of the tool. Upper Twin Mountain, in contrast, shows a distinctive pattern with retouch most common distally and gradually decreasing in frequency clockwise around tool edges, with the lowest frequencies occurring on the left edges of tools. This pattern could be explained by site function, since a much more limited range of activities would be expected to have occurred at a kill site as opposed to a camp. An alternative explanation is that this pattern is a result of a small sample size, with only five tools having been recovered from Upper Twin Mountain. Nevertheless, this differentiation between the camps and kill reinforces the notion that technological variability is in part governed by site function, a pattern that recurs regularly throughout this study.

TABLE 7.5. Percentage of tool edge units exhibiting retouch by site.

| Site | | EU1 | EU2 | EU3 | EU4 | EU5 | EU6 | EU7 | EU8 | EU9 | EU10 | Sum |
|---|---|---|---|---|---|---|---|---|---|---|---|---|
| CKM | Used | 9 | 11 | 12 | 11 | 8 | 5 | 6 | 9 | 13 | 11 | 95 |
| | Available | 13 | 15 | 16 | 16 | 12 | 10 | 12 | 16 | 16 | 16 | 142 |
| | Percent | 69.2% | 73.3% | 75.0% | 68.8% | 66.7% | 50.0% | 50.0% | 56.3% | 81.3% | 68.8% | 66.9% |
| AGB | Used | 46 | 55 | 52 | 48 | 48 | 25 | 39 | 51 | 51 | 59 | 474 |
| | Available | 61 | 70 | 76 | 79 | 73 | 66 | 75 | 80 | 78 | 78 | 736 |
| | Percent | 75.4% | 78.6% | 68.4% | 60.8% | 65.8% | 37.9% | 52.0% | 63.8% | 65.4% | 75.6% | 64.4% |
| KRMP | Used | 37 | 37 | 36 | 31 | 24 | 18 | 33 | 35 | 43 | 42 | 336 |
| | Available | 57 | 59 | 63 | 60 | 49 | 48 | 59 | 64 | 69 | 61 | 589 |
| | Percent | 64.9% | 62.7% | 57.1% | 51.7% | 49.0% | 37.5% | 55.9% | 54.7% | 62.3% | 68.9% | 57.0% |
| BGB | Used | 69 | 72 | 63 | 53 | 27 | 9 | 34 | 50 | 58 | 66 | 501 |
| | Available | 114 | 117 | 136 | 132 | 114 | 99 | 114 | 130 | 131 | 128 | 1,215 |
| | Percent | 60.5% | 61.5% | 46.3% | 40.2% | 23.7% | 9.1% | 29.8% | 38.5% | 44.3% | 51.6% | 41.2% |
| UTM | Used | 3 | 4 | 3 | 3 | 2 | 1 | 1 | 1 | 1 | 1 | 20 |
| | Available | 4 | 5 | 5 | 5 | 4 | 4 | 4 | 5 | 5 | 5 | 46 |
| | Percent | 75.0% | 80.0% | 60.0% | 60.0% | 50.0% | 25.0% | 25.0% | 20.0% | 20.0% | 20.0% | 43.5% |

TABLE 7.6. Measures of average tool use-life by site.

| | Site | | | | |
|---|---|---|---|---|---|
| Variable | CKM | AGB | KRMP | BGB | UTM |
| Average retouch intensity | 2.14 | 1.80 | 1.75 | 1.42 | 1.60 |
| Average edge:tool thickness | 0.63 | 0.52 | 0.52 | 0.37 | 0.36 |
| % edge units retouched | 66.9 | 64.4 | 57.0 | 41.2 | 43.5 |

Returning to the problem at hand, to test the hypothesis that mean tool use-life declines with longer occupation spans (and therefore greater raw material availability), I combined the three variables representing retouch intensity into a single index, the retouch intensity index, using a method identical to that used for the core reduction index. I transformed each variable to scale from 0 to 1 by linear interpolation between the maximum and minimum values. I then averaged the standardized values for each variable by site (table 7.6). Figure 7.6 shows the relationship between the occupation span and retouch intensity for the five sites. Again, Upper Twin Mountain stands as an outlier. When Upper Twin Mountain is excluded from the analysis, there is strong support for the model. The natural logarithm of the OSI is significantly negatively correlated with the retouch intensity index ($r = -0.979$, $p = .012$). If the average degree of retouch present on tools is an indicator of tool use-life, then average tool use-life declines with longer occupations.

While this analysis supports the notion that tool use-life responds to occupation span, the marginal value theorem predicts only that it should be a function of raw material availability. As I have argued above, local raw materials are available at all of the sites, and therefore the natural raw material availability should be relatively similar among all five sites, with the possible exception of Barger Gulch (see chap. 5). Therefore, cultural raw material availability, or the amount of stone stockpiled, should correlate well with average tool use-life. To test this proposition, I repeated the above analysis using the core mass density (g of stone per m$^2$) in lieu of the OSI (fig. 7.7). Core mass density is a measure of the amount of stone available for tool production. Upper Twin Mountain again breaks from the general trends, and when it is excluded the natural log of core mass density is significantly negatively correlated with the retouch intensity index ($r = -0.984$, $p = 0.016$).

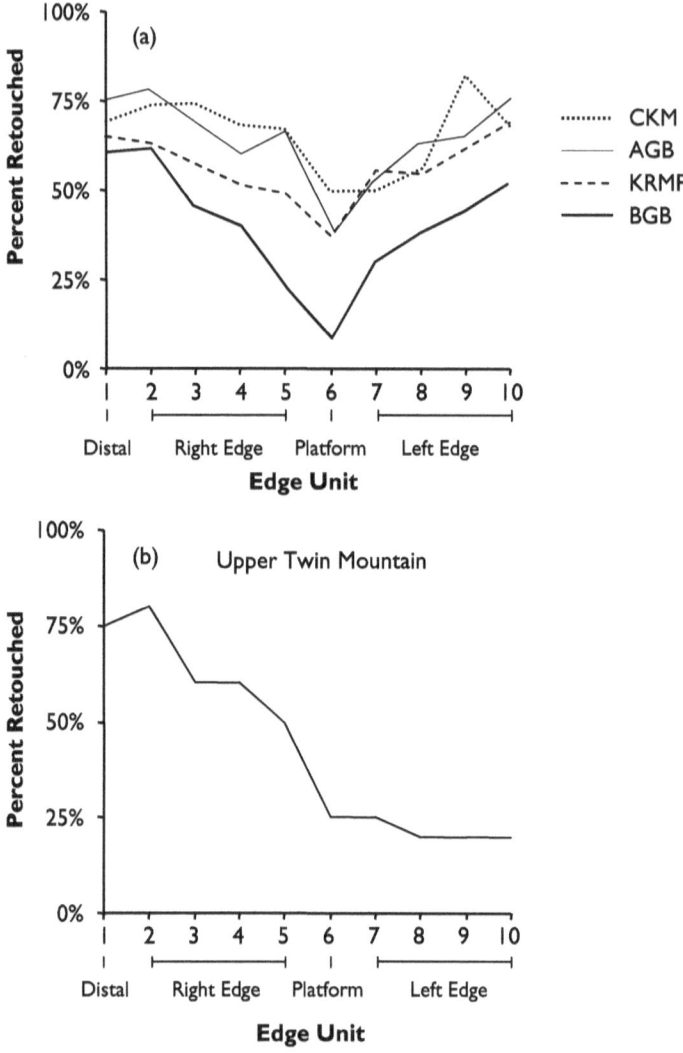

FIGURE 7.5. (a) The percentage of retouch on flake tools by edge unit for the four campsites. (b) Same for Upper Twin Mountain.

## Discussion

In this section I explored the relationship between occupation span, raw material availability, and flake tool use-life as reflected by the intensity of retouch. Using the marginal value theorem, I predicted that average tool use-life should decrease as occupation span and raw material availability

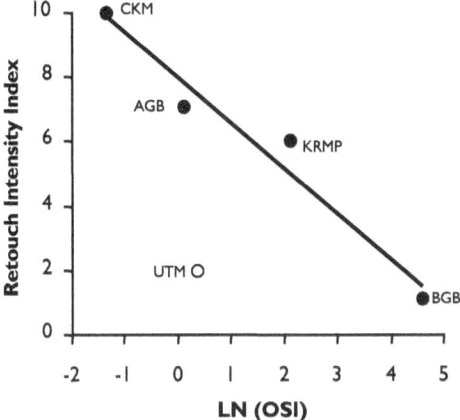

FIGURE 7.6. Linear regressions of the Ln (occupation span index) vs. the retouch intensity index. Regression excludes Upper Twin Mountain.

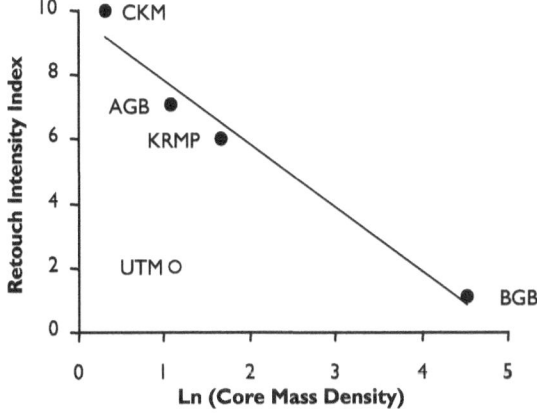

FIGURE 7.7. Linear regression of the Ln (core mass density [g/m²]) vs. retouch intensity index. Upper Twin Mountain (open circle) excluded.

increase. The data from the four campsites in the study sample provide support for this hypothesis. In short-term occupations, therefore, there is an emphasis on tool maintenance, the repeated resharpening of tools already on hand. As a critical mass of tool stone is built up in a site, I argue that the degree of tool maintenance declines as the emphasis shifts to the rapid replacement of tools.

This phenomenon alone could explain the problem that initiated this analysis—increasing frequencies of core reduction flakes relative to bifacial thinning flakes as occupations are prolonged (fig. 7.1). From the null model of debitage discard (equation 7.6), if tool use-lives decrease as a function of occupation span, then the discard rate of core reduction flakes should correspondingly increase since tools will be manufactured at shorter and shorter intervals. Of the three hypotheses presented in figure 7.2, this analysis would suggest that either hypothesis 1 or 3 is correct. Because hypothesis 2 predicts no change in the discard rate of core reduction flakes, it is provisionally (and circumstantially) falsified.

## How Many Flakes Does It Take to Make a Tool?

In this section, I explore three additional components of the null model: the number of flakes produced per flake tool manufacture ($f_t$), biface manufacture ($f_{bp}$), and biface maintenance ($f_{br}$) events. Again, changes in any of these variables could potentially produce the pattern in figure 7.1. Presumably, though, to produce the observed pattern, over the course of an occupation, more core reduction flakes would have been produced per tool manufacture event and/or fewer bifacial thinning flakes would have been produced per biface manufacture or maintenance event. I begin the discussion with flake tools.

Borrowing the title of this section, how many flakes does it take to produce a flake tool? The simplest answer is that minimally it takes one. You can, for example, use the first flake removed as a tool blank. However, that blank may not have the most desirable properties for the type of tool you want to manufacture. So, you might make further removals until you get a flake close to the ideal form you had in mind, or you may remove a number of flakes in preparation of the core to adapt the core face so as to facilitate the production of the desired form as in Levallois reduction. Binford and O'Connell's (1986, 1989) accounts of Alyawara flintknappers provide examples of both behaviors (for archaeological examples, see Olszewski 1989; Moss 1986):

> It was clear from watching Sandy that he knew the type of flake he wanted to produce. It was also clear that he knew he had to alter the core's shape and then strike its edge relative to the shape of the face in order to produce a flake of desired form. (Binford and O'Connell 1989:127)

The blanks that the men wanted to save for particular manufacturing needs were generally placed near the right knee of the worker and picked up when the worker abandoned the core. (Binford and O'Connell 1986:453)

Presumably, the blanks not so lucky as to receive the "right knee" placement were simply discarded. On average then, it could take potentially very few flakes to produce a flake tool if a knapper is not particularly picky or if a core is already set up for a desired removal, or it could take many flakes to produce a tool if the opposite conditions hold. Flintknapping skill is another variable that could certainly also come into play. In this analysis, I focus only on the variable selectivity, or how picky a flintknapper is in selecting flake blanks to be used for tools, although core preparation and skill could be modeled similarly.

Following Kuhn (1994:438–439), I assume that for a given type of tool, there is an ideal flake blank. In this case, by "ideal" I mean the blank form that maximizes functional efficiency, or the average work performed per unit time over the use-life of a tool. That blank could be described by any number of variables, but for simplicity I focus only on size. Any deviation from the optimally sized tool blank would yield a loss of functional efficiency. We could, therefore, describe a function relating size to functional efficiency. Further assume that core reduction produces some size distribution of flakes. These assumptions are depicted graphically in figure 7.8a. Using these two functions, it is possible to derive the probability of producing a flake of any given value of functional efficiency. More directly for the question at hand, how does the number of flakes produced relate to the average maximum functional efficiency you can hope to produce from a set of n flakes? In other words, how do your chances of producing a flake, which exceeds all of the previous flakes produced, change as a function of the number of flakes removed?

Because of the complexity of calculation in this problem, a simple simulation of the process can be used to derive the general relationship. In this simulation, random flakes are produced within a given size distribution, and functional efficiency is assumed to be a normal function of artifact size approximating the functions shown in figure 7.8a. The average maximum functional efficiency is calculated for producing n flake blanks over one thousand iterations. As shown in figure 7.8b, as more

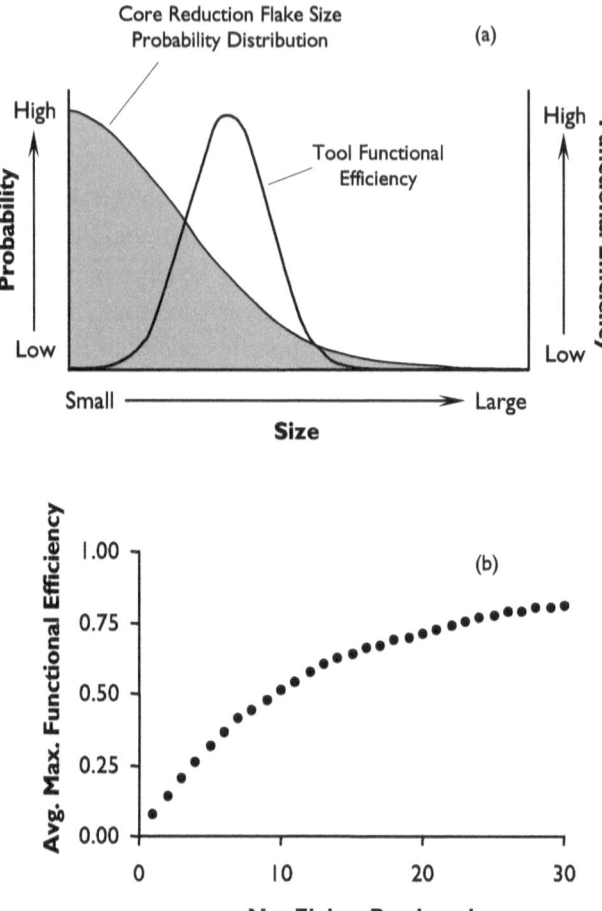

FIGURE 7.8. (a) A conceptual model of flake tool design and flakes produced in core reduction. The gray area shows the probability distribution of flake production with respect to size. The normal curve represents the relationship between function efficiency and size for a flake tool. (b) The number of flakes produced vs. the average maximum functional efficiency for that set of flakes. Each point in the curve was generated by 1,000 iterations (selecting $n$ flakes in each iteration) of a simulation assuming that tool functional efficiency was described by a normal function where $\mu = 30$, $\sigma = 4$, multiplied by 10 to standardize efficiency values to range from 0 to 1. Flakes were produced of random size within a normal distribution ($\mu = 0$, $\sigma = 15$), allowing only positive values.

flakes are produced, the average maximum blank functional efficiency increases toward an asymptote that describes the maximum potential functional efficiency for a given tool type. If there is no cost to producing hundreds of flakes, then a knapper should simply keep producing flakes until an optimal blank is manufactured. But there are costs to such behavior. Reducing all of your cores to rubble in search of the perfect flake limits your potential to produce future flake blanks and requires the procurement of additional raw material for future tool production. Therefore, there is a raw material opportunity cost to being picky. This is an opportunity cost, since as the number of flakes produced increases, your ability to control future reduction decreases.

Following this idea, a simple model shows that there should be an optimal number of flakes produced per tool to maximize benefits and minimize raw material costs. Assume a knapper wants to maximize production efficiency, defined as the average maximum blank functional efficiency minus raw material costs. For simplicity, it is assumed that raw material costs increase as a linear function of the number of blanks produced and raw material availability. From figure 7.9a, production efficiency increases to an optimum and declines as additional flakes are produced. In this example, the optimum number of flakes to produce per tool is fourteen. The model also predicts that the optimum number of flakes to produce per tool increases as raw material availability increases (fig. 7.9b). In other words, having a lot of raw material on hand and/or being located near a natural source of lithic raw material allows a knapper to be choosy in selecting flake blanks to maximize tool utility. The exact form of the relationship between raw material availability and how many flakes should be produced per tool will depend on the functions plugged into the model. Therefore, as I have loosely constructed the model, it is only expected that as occupations are prolonged and raw material becomes increasingly available, flintknappers should have been increasingly picky in tool production and produced, on average, more flakes per tool manufactured (see Rolland and Dibble 1990).

Before testing the model, I return briefly to bifaces. The number of bifacial thinning flakes produced per biface manufacture and maintenance events should have little, if anything, to do with choosiness. In large part, it should be governed by the size of the biface manufactured or reduced. Larger bifaces with more edge will produce more bifacial thinning flakes

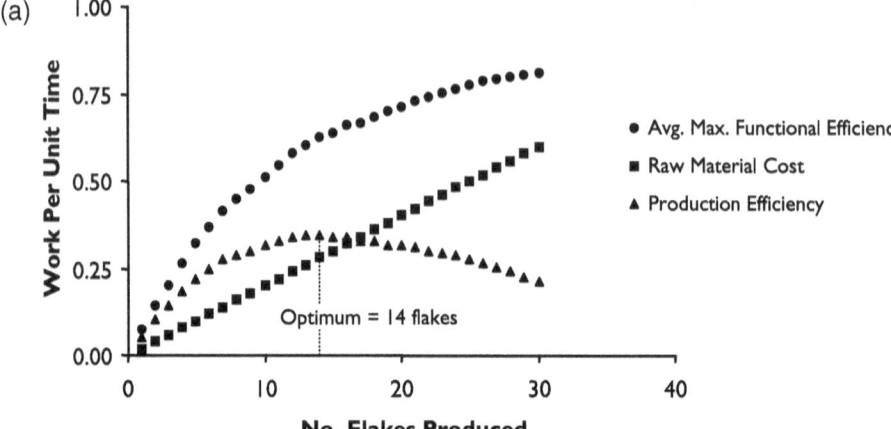
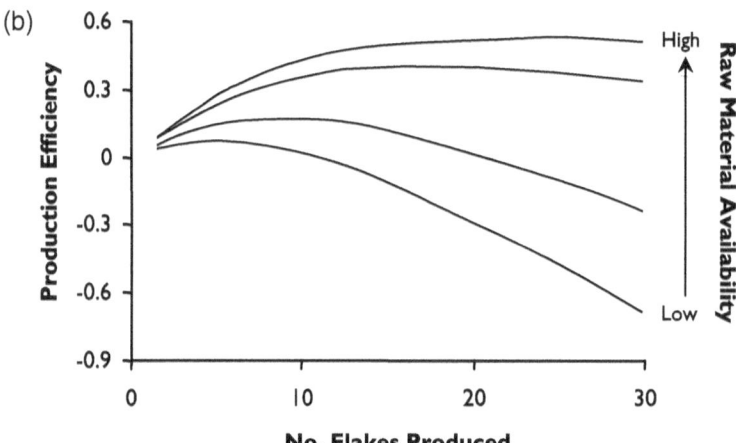

FIGURE 7.9. (a) The number of flakes produced in tool manufacture vs. the average work gained or lost per unit time for the average maximum functional efficiency of flake blanks, raw material costs, and the production efficiency (gain-loss). The curves were determined by 1,000 iterations of a simulation assuming that tool functional efficiency was described by a normal function of size ($\mu = 30$, $\sigma = 4$) multiplied by 10 to standardize from 0 to 1. Flakes were produced of random size within a normal distribution ($\mu = 0$, $\sigma = 15$), allowing only positive values. (b) The number of flakes produced vs. production efficiency with varying raw material availability. As raw material becomes increasingly available, the optimum number of flakes to produce per tool manufacture episode increases.

when an edge is renewed or when that biface is first produced. Unless the average size of bifaces in the toolkit changes as a function of occupation span, the average number of bifacial thinning flakes produced per event should remain constant. Because bifaces are constantly being maintained, discarded, and manufactured at various stages in their use-lives, it would be expected that the average size of bifaces in the toolkit will vary little and, therefore, so should the rate of bifacial thinning flake discard.

*Testing the Model*

Two predictions were developed. First, flintknappers should be increasingly choosy in tool blank production with increased raw material availability. Because raw material availability increases with occupation span (fig. 7.3), there should be a correlation between selectivity and occupation span. Selectivity can be monitored by the ratio of core reduction flakes to flake tools. If very few waste flakes are produced per tool, the ratio of flakes to tools should be low; if many waste flakes are produced per tool, the ratio of flakes to tools should be high. The second prediction was that occupation span and raw material availability should have no effect on the number of bifacial thinning flakes per manufacture and maintenance events, and, therefore, the ratio of bifacial thinning flakes to bifaces should remain constant.

To avoid artificial inflation of debitage samples for sites with small assemblages, I limit my analysis to include only debitage larger than 2 cm in maximum dimension. Unfortunately, this seriously limits sample size for Carter/Kerr-McGee and Upper Twin Mountain, but given the design of my analysis, this is unavoidable. Counts of debitage, flake tools, and bifaces are presented in table 7.7, as are the ratios of core reduction flakes to flake tools and bifacial thinning flakes to bifaces.

Figure 7.10a shows the relationship between core mass density and the ratio of core reduction flakes to flake tools. A significant positive log-linear correlation is evident for the four campsites, providing support for the model ($r = 0.953$, $p = 0.047$). In contrast, there is a weak positive correlation (log-linear) between core mass density and the ratio of bifacial thinning flakes to bifaces, but it is not statistically significant (fig. 7.10b; $r = 0.565$, $p = 0.435$). At Carter/Kerr-McGee, there are four bifacial thinning flakes per biface, while at the other three campsites there are

TABLE 7.7. Counts of debitage,[†] tools, and bifaces by site.

| Variable | Site | | | | |
|---|---|---|---|---|---|
| | CKM | AGB | KRMP | BGB | UTM |
| Core reduction flakes | 5 | 161 | 252 | 864 | 5 |
| Flake tools | 18 | 85 | 72 | 153 | 5 |
| CRF:flake tools | 0.28 | 1.89 | 3.50 | 5.65 | 1.00 |
| Bifacial thinning flakes | 12 | 177 | 98 | 198 | 0 |
| Nonprojectile point bifaces | 0 | 7 | 10 | 14 | 0 |
| Points and preforms | 3 | 20 | 6 | 17 | 2 |
| Total bifaces | 3 | 27 | 16 | 31 | 2 |
| BFT:bifaces | 4.00 | 6.56 | 6.13 | 6.39 | 0.00 |

Note: [†]Sample limited to debitage larger than 2 cm.

approximately six. With these data alone, it is difficult to reject the null hypothesis that there is no correlation between occupation span and the bifacial thinning flake:biface ratio, but given the comparative strength of other correlations performed in this study, this analysis is certainly suggestive of duration of occupation having little, if any, impact on the ratio. One of the three bifaces from Carter/Kerr-McGee is a very small (length = 7 mm) unground ear of a Folsom preform, which likely does not represent a failed biface, and if removed from the analysis, it would increase the ratio to 7.0, in line with the other campsites.

A chi-square test was performed to further evaluate the null hypothesis that the relative frequencies of bifaces and bifacial thinning flakes (table 7.7) are constant among the campsites of the study sample. Again, no significant difference was found ($\chi^2 = 0.56$; df = 3; p = 0.91), suggesting that the ratio of bifacial thinning flakes to bifaces does not vary among the campsites of the study sample. A comparable chi-square test using the frequencies of core reduction flakes and flake tools was highly significant ($\chi^2 = 96.49$; df = 3; $p \ll 0.001$).

## Discussion

It was argued that abundance of raw material affords the luxury of being wasteful. When lithic raw material is scarce, I propose that functional efficiency of tools is sacrificed to minimize raw material costs. Changes in

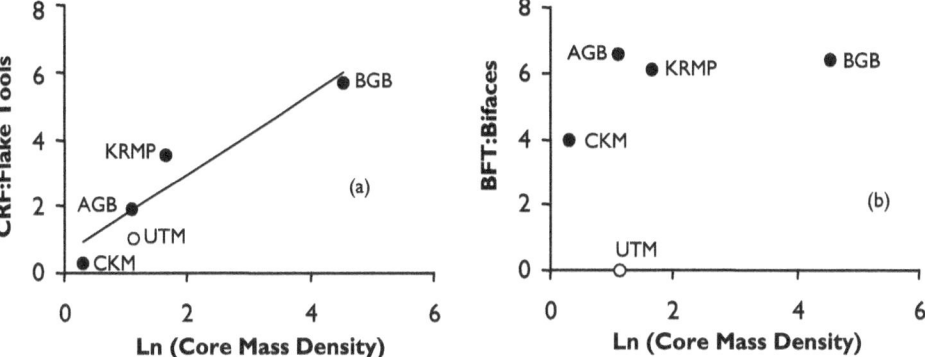

FIGURE 7.10. (a) Linear regression of Ln (core mass density [g/m²]) vs. the ratio of core reduction flakes to flake tools. Regression excludes Upper Twin Mountain (shown as open circle). (b) Scatterplot of the mobility index and log (mobility index) vs. the ratio of bifacial thinning flakes to bifaces.

the ratio of core reduction flakes to flake tools provide support for this proposition. At sites characterized by long-term occupations and stockpiled raw material, it appears that flintknappers were increasingly picky when it came to selection of flake blanks for tool forms. So how many flakes does it take to make a tool? If raw material is abundant, it takes many. If raw material is scarce, it takes few. I should caution, though, that there are other possible explanations of this pattern. For example, as occupations are prolonged, it is possible that a greater proportion of flake tools are discarded off-site. This might be expected since long logistical forays are made possible and expected with extended stays in a single location (Kelly 1983; Kuhn 1989, 1995; Sahlins 1972; Surovell 2000), and the average number of tools discarded per foray should be proportional to the length of the trip. Another possibility is that as occupations are prolonged a greater proportion of tools are used and discarded without retouch (Parry and Kelly 1987), and therefore would not be recognized as tools in this study. The latter idea is consistent with the analysis of tool use-life performed above because unretouched tools recovered archaeologically were presumably discarded very early in their potential life span. In contrast to flake tools, it was argued that the number of bifacial thinning flakes produced per biface manufacture and maintenance events should be largely a function of biface size, and therefore should bear no relationship to occupation duration. Modest support for this hypothesis was found.

Combining these two observations allows further evaluation of the three hypotheses (fig. 7.2) presented earlier to explain changing debitage frequencies across the sites of the study sample (fig. 7.1). This analysis solely supports hypothesis 1—the rate of discard core reduction flakes increases with occupation span, while the rate of discard of bifacial thinning flakes remains constant. With these data alone, it is difficult to reject hypothesis 3, that the discard rate of bifacial thinning flakes also decreases over the course of an occupation. If hypothesis 3 is correct, however, the analysis suggests three possibilities: (1) fewer bifaces are maintained in the toolkit, (2) the use-lives of bifaces shorten, and/or (3) the use-lives of bifacial edges shorten as occupations are prolonged.

I began this analysis with an accumulation equation (7.6), in which I modeled the discard of debitage as a by-product of flake tool production and biface production and reduction. The model predicted that the discard rates of core reduction and bifacial thinning flakes should remain constant through the course of an occupation. An analysis of debitage frequencies showed, in contrast, that core reduction flakes become increasingly dominant in assemblages with longer occupation spans (fig. 7.1). This led to three hypotheses (fig. 7.2) concerning changes in discard rates of these debitage types over time: (1) the discard rate of core reduction flakes increases with longer occupation spans, (2) the discard rate of bifacial thinning flakes decreases with longer occupation spans, or (3) both occur simultaneously. Using the null model as a guide, indirect tests of these hypotheses were performed. I argued first that the variable driving the system was not occupation span but instead changes in raw material availability that correlate with occupation span. Using the marginal value theorem, I argued that the average use-lives of tools should decrease with greater raw material availability, a hypothesis that was supported using three measures of retouch intensity (figs. 7.6 and 7.7). I then developed a simple model of selectivity of flake blanks in tool manufacture and argued that with greater raw material availability, flintknappers should be increasingly picky in choosing which flakes to use as tool blanks. Changing ratios of core reduction flakes to flake tools provided support for this hypothesis as well (fig. 7.10a). In contrast, no correlation was found between the ratio of bifacial thinning flakes to bifaces (fig. 7.10b). These analyses provided circumstantial support for hypothesis 1. As tool use-lives decrease, and tools are replaced more frequently, the rate of core reduction flakes increases (equation 7.6). Likewise, as flintknappers become increasingly choosy in tool manufacture, the rate of core reduction flakes should further increase. Therefore, in short-term occupations with little raw material stockpiled, flake tools are heavily maintained and toolmakers sacrifice functional efficiency for stone conservation.

The fundamental arguments about intersite variability and behavior presented in this analysis are not new. Parry and Kelly (1987) and Kuhn (1994), for example, argued that changes in raw material availability caused by the stockpiling of stone in long-term occupations have a ripple effect, resulting in modifications of technological behaviors. Parry and Kelly (1987) suggest that with reduced residential mobility comes a shift to informal tools that are more efficient in use, an argument very similar to my application of the marginal value theorem. Kuhn (1994:33–34) likewise suggests that as occupations are prolonged and places are provisioned, tools should be discarded earlier in their use-lives. It has also been argued that raw material availability should affect flake blank selection as I have argued (Rolland and Dibble 1990:485). The value of this analysis is not that it is novel in substance, but in form. I have shown how the discard of debitage, unwanted by-products of tool production, can be modeled as an epiphenomenon of the optimization of tool production, use, and discard, and the use of formal models in this study allowed the construction of clear, unambiguous links between changes in tool use-lives and selectivity in tool blanks and its consequences for changing debitage frequencies.

The specific problem examined in this study is a classic North American problem—what controls the frequencies of bifacial and nonbifacial (or formal and informal) artifacts and their by-products in archaeological sites (Andrefsky 1991, 1994; Bamforth and Becker 2000; Cowan 1999; Johnson 1989; Parry and Kelly 1987). Although I did not examine the relationships between bifaces, cores, and flake tools directly, I did establish their roles and presence or absence in mobile and immobile toolkits in the previous chapter, and I examined the by-products of their production and reduction in this chapter. I do not feel comfortable generalizing beyond the sites of the study sample but do propose that for Folsom and Goshen hunter-gatherers of the Northwestern Plains and Rocky Mountains, a number of currencies are governing the system. In the previous chapter, I argued that bifaces were designed to maximize transport efficiency and were transported over long distances. Flake tools also were components of mobile toolkits but were designed largely to meet functional requirements. Core reduction, geared toward flake tool production, was limited to sites in the proximity of natural sources of lithic raw materials and cores were not transported great distances.

With these considerations, high frequencies of bifacial thinning debris in short-term occupations, however, do not mean that bifaces were necessarily favored for transport over flake tools in contexts of high mobility. Bifaces were certainly favored over the potential to produce flake tools (i.e., cores), but both bifaces and flake tools were regularly transported. Instead, I suggest that bifaces and flake tools were a constant; they were always present in toolkits. In contrast to flake tools, however, the rates of biface and bifacial thinning flake discard likely remained relatively constant. What is largely governing the frequencies of these artifacts types and their by-products are differences in flake tool production, governed largely by raw material availability. When raw material is abundant, flake tools and core reduction flakes are discarded at increased rates, resulting in their numerical dominance in assemblages.

## Raw Material Availability

In this chapter I have also tried to redefine the concept of raw material availability from simply a natural property of locations to a variable that is also governed by human behavior, as many individuals have suggested (e.g., Bamforth 1990a,b; Kuhn 1991, 1994; Nelson 1991; Parry and Kelly 1987). From the perspective of a location, raw material availability is a function not only of the amount of stone available at a site in natural and cultural sources, but also of the rate at which stone is used and the rate at which you are approaching a natural source of stone. For all the sites of the study sample, natural sources of stone are immediately available, which allowed the use of the amount of stone available in cores as a proxy measure of raw material availability. It was shown that raw material availability and occupation are indeed positively correlated. I did not examine the effects of the rate of consumption of stone or the rate of mobility, but it is interesting to note that the faster one moves across a landscape toward a source of lithic raw material (frequent residential mobility), all things being equal, your intrinsic raw availability will be higher. On the other hand, the longer you stay in one place (infrequent residential mobility), your potential to stockpile stone is enhanced, highlighting the need for rethinking the concept of lithic raw material availability and pointing to directions for additional modeling and archaeological tests. Raw material availability is far more than distance to source.

*Surpluses Reconsidered*

In chapter 5, I argued that lithic surpluses were created to cushion against periodic shortfalls that would require extra raw material procurement trips. In this chapter, I have shown that surplus creation correlates with enhanced rates of core consumption in flake tool production. More lithic raw material stockpiled allows greater tool efficiency through careful selection of tool blanks and replacement of tools early in their use-lives. So are surpluses created and maintained to prevent shortfalls or to enhance tool efficiency? The answer is likely that surpluses are created and maintained for both reasons, which may in part explain why the observed surpluses in the sites of the study sample are so large (fig. 5.10).

## A Spherical Cow Rears Its Rounded Head

In chapter 1, I introduced what I called the "spherical cow problem," the construction of informal models with invalid starting assumptions. This chapter presents a clear contradiction in assumptions between models. In chapter 3, in developing the OSI, I assumed that artifact discard rates remained constant over the span of an occupation. In this chapter, I have shown that the discard rate of core reduction flakes likely increases as occupations are prolonged. This is not necessarily a contradiction. If the discard rate of some other component of assemblages correspondingly decreases as the discard rate of core reduction flakes increases, then the average discard rate for all artifacts could remain constant. Although plausible, I find this unlikely. I suspect that artifact discard rates in general increase with longer occupation spans and eventually plateau.

Have I shot myself in the foot? I have used the OSI throughout this study as a variable against which I have monitored technological variability. Nonlinearity in the function relating discard rates to occupation span will affect only the exact form of correlations between the OSI and independent technological variables and does not invalidate the OSI as a measure of mean per capita occupation span.

# 8

# Mathematics, Lithic Technology, and Paleoindians

> A good model is a clear model, one that can be understood and used to produce the same results by any interested investigator. . . . A good model provides the same answers to all comers.
> —John Casti, *Would-Be Worlds*, 1997

THIS STUDY IS intended foremost to be an exercise in the application of formal optimality models to the study of lithic technology. It seems hardly necessary at this juncture to reiterate the values of models to archaeological endeavors, but I do so to emphasize that rigor in model construction paves the way for clear explanation. In my opinion, one of the most elegant statements on the value of models to archaeology was written by David Clarke and appears on the third page of the 1,055-page edited volume *Models in Archaeology* (Clarke 1972a). I repeat much of it here:

Why need the archaeologist concern himself with models? There are five main reasons, which may be briefly outlined:

(1) Whether we appreciate it our not, our personal archaeological opinions, approach, aims and selection of projects are controlled by largely subconscious mind models which we accumulate through time. We should realize that we are thus controlled.

(2) Whether we appreciate it or not, we always operate conceptual models in the interpretation of observations. We all resemble the Molière character who was delighted to find that all his life, unknowingly, he had been speaking prose. We should make these operational models explicit and testable.

(3) The construction, testing, verification or refutation of modification of explicit models is the essence of empirical and scientific approaches.

(4) The existence of a model presupposes the existence of an underlying theory, since a model is but one simplified, formalized and

skeletal expression of a theory—be it tacit or explicit-developed for a particular situation.

(5) Finally: Hypotheses are generated from the model expression of a theory. Explanation comes from tested hypotheses. Hypotheses are tested by using relevant analyses on meaningful categories of data.

Thus models are a vital element in all archaeological attempts at hypothesis, theory, explanation, experiment, and classification. (Clarke 1972b:3)

The first two points are particularly relevant to the use of formal models. Not long ago during a summer field season in Colorado, I was given a tour of a handful of archaeological sites by a gentleman who was kind enough to share his perspective on local archaeology and prehistoric human adaptations. He was not a professional archaeologist, but he loved prehistory and was delighted to have the chance to share his opinions with a student of archaeology. In one instance he was explaining where Paleoindian bison kill sites should be found based on the way that he would have hunted bison if given the chance 10,000 years prior. He began this mini-lecture with the statement, "If I was an abo[riginal] . . ." As much as I believe we have moved beyond this form of model building in American archaeology, it is, without pointing fingers, still surprisingly common. Operating in this mode, the "projection of ourselves into the past," or what I have heard referred to as "egocentric analogy," we run the risks of (1) not learning anything beyond our preconceptions, and (2) building models based on nothing more than wisdom of our past experience, which has no guarantee of being good wisdom or of having logical rigor. Often, informal qualitative models take this form, though rarely do they do so explicitly. This is not to say that formal models are always good descriptions of the world, or that knowledge cannot be gained through induction. I only wish to make the point that common sense approaches to understanding optimal choices can be just plain wrong.

An excellent example of the inability of the human brain to fully grasp optimal decision making using commonsense reasoning is the Monty Hall problem. This problem was first introduced to me by my PhD advisor, Steven Kuhn, at the University of Arizona, but it was popularized when it appeared in an "Ask Marilyn" column in *Parade* magazine on September 9, 1990. The problem is commonly stated this way:

> You are a participant in a game show. Three doors are placed in front of you. Behind one door is the $10,000 grand prize. The host asks you to select a door. You choose door No. 2. The host then opens door No. 3 to reveal 40 quarts of used motor oil. Next, the host gives you the choice to switch to door No. 1 or to stick with your original choice. What should you do? To maximize your probability of winning the grand prize, should you stay with your original choice, should you switch doors, or does it not matter?

If you want a moment to think about this problem and create a "subconscious mind model" in the words of David Clarke, I suggest you do it right now, because I am about to reveal the answer. If you are unfamiliar with this problem, you probably think, like most people, that it does not matter whether you switch or not. Your reasoning goes something like this: Since the host removed one door from consideration and two doors now remain, your chance of winning the grand prize is 50 percent, and this probability remains constant whether you switch doors or not. In short, it does not matter; by switching doors you do not increase or decrease your chances of winning.

You may be surprised to discover that this is not correct. In fact, you double your chances of winning by switching doors, and it is fairly simple to demonstrate that this is true using the mathematical laws of probability. No doubt you will agree that when you made your original choice, the probability of guessing the correct door, the door concealing the grand prize, was 1 in 3. You will then also agree that your chance of guessing an incorrect door was 2 in 3. What you might have a hard time accepting is that these probabilities remain unchanged when the host reveals that the grand prize is not behind door No. 3. By switching doors, your chance of winning increases to 2 in 3. By staying with your original choice, your probability of winning remains 1 in 3. The optimal strategy (assuming a goal of winning the grand prize, under the constraints that the host never reveals the position of the grand prize when he opens a door, and always gives you the option to switch doors) is to switch doors. Here is one way to conceptualize this problem. If you chose incorrectly in your first choice of doors, and you decide to switch, you will always win the grand prize. Your probability of choosing incorrectly in your initial choice is 2 in 3. Thus, by switching doors every time, your chances of winning are 2 in 3.

I have posed this problem to students in a quantitative methods course I teach, and every year I have students who absolutely refuse to accept that this solution is correct. So, you are not alone if you do not believe me. When this problem was originally posed to me, I had a difficult time accepting this answer until I sat down with three coffee mugs with one concealing a penny and reenacted it numerous times. This only serves to demonstrate my point—our common sense can easily fail us when we are trying to figure out what goal-oriented humans should choose to do when facing a decision. Although this problem may have little direct applicability to archaeological situations, there are somewhat counterintuitive predictions of formal models that have direct archaeological implications. Here is an example.

Given a goal of minimizing walking distances while foraging in a homogenous environment, should hunter-gatherers choose to be central-place foragers or not? Because all ethnographically known hunter-gatherers are central-place foragers, many people might assume the answer to this question to be "yes," but the answer is "no," and it is fairly simple to demonstrate this mathematically (Surovell 2000). It is much more efficient to move from patch to patch rather than from a camp to a patch and back to a camp again. This likely explains why central-place foragers are the minority in the animal kingdom. It also explains why we occasionally abandon central-place foraging. When we forage in grocery stores, for example, we grab a cart or basket and move from aisle to aisle collecting our goods. We do not place the cart in the middle of the store and then make round-trips from our cart to the meat, the dairy, and produce sections, because doing so would dramatically increase our walking distance and shopping time. Finally, this simple and somewhat counterintuitive implication of foraging models raises the important question of why humans became central-place foragers in the first place.

Returning to the Monty Hall problem, there is one more important point to be made about my personal experience in accepting the answer. To convince myself that I should always switch doors to maximize my chance of winning, I created a model involving three coffee cups and a penny. Like all models, my model was an abstraction and simplification of the problem. Coffee cups are not doors. A penny is not $10,000. And unfortunately Monty Hall did not come to my house. Nonetheless, this abstract representation of the system I was trying to understand served its purpose: it demonstrated the implications or predictions of a series of assumptions. The model was explicit in its decision variable, constraints, and currency, though it was not formal. What is interesting about my coffee cup model is that it convinced me of the correct answer only after I knew what the correct answer was. Formal mathematical models, like coffee cups and pennies, are simplified abstractions, but unlike with informal models, your prior knowledge plays no part in their predictions, except of course that your preconceptions may shape the assumptions built into a model. No matter how good you think your commonsense understanding of something may be, formalizing a system as a mathematical model can be a humbling experience, demonstrating very quickly that your common sense is not as sensible as you thought it was.

At this point, you might raise the objection that if human common sense can be logically flawed, then why should I accept the premise that mathematical models of human behavior are good descriptions of human behavior? I would answer this objection in two ways. First, a model is not reality. An optimality model says what people should do. What people actually do is a separate issue, which is why, of course, a tested model is vastly superior to one that stands alone. Secondly, my hunch is that given a problem that humans face repeatedly, particularly a problem with serious consequences, people will, in short order, determine optimal solutions by the process of trial and error. I would not be surprised, for example, if actual game show contestants on *Let's Make a Deal* did not conform to the predictions of the Monty Hall optimality model because this is a problem each contestant faced only once or twice. If, however, professional *Let's Make a Deal* players existed, people who faced this problem over and over again, I suspect that they would quickly conform to the optimal solution. In this sense, the people we study in the archaeological record were almost certainly of the "professional" rather than the "amateur" variety of decision makers, and no matter how complex a problem was, they would have found the optimal solution in short order. For example, for a rookie flintknapper there is relatively little guidance from common sense concerning the optimal angle to strike a core to produce a flake, but with minimal trial and error, this problem is easily solved.

The point is that we necessarily bring our personal experience to archaeological model building and interpretation, and although explicitly spelling out models in mathematical terms does not eliminate personal biases, it does mean that models are explicit, unambiguous, logically sound, and have undeniable predictions. Relying on commonsense perceptions can lead us astray. This is not to say that because a model is constructed mathematically, it is a good description of the world. I can create, for example, a mathematical model that predicts that an apple released from your hand on Earth on a still day will fall upward. Such a model would have faulty assumptions. We can question whether the assumptions built into a formal model are correct and whether they apply to a particular application of that model, but we cannot question whether the predictions of that model follow from its assumptions. Furthermore, when we construct mathematical models, they often have counterintuitive predictions, sometimes predictions that are opposite of those we anticipate.

The use of formal mathematical models is increasingly common in other fields that have a tradition of qualitative model building, most notably biology. In a paper titled "Biologists Put on Mathematical Glasses," Fagerström et al. (1996) suggest that the neo-Darwinian synthesis in biology, linking classic Darwinian evolution with sophisticated formal models of population genetics, represented "the first case of a reasonably mature, proper theoretical biology." This statement should, to some degree, make archaeologists cringe. If expressing models as formal mathematical abstractions is the foundation of a "mature, proper theoretical" science, then one could reasonably question whether archaeology has achieved that status. In other words, are we immature as a science because we have not made the leap from largely qualitative to quantitative explanations of the operation of the prehistoric world? Of course, we could reject the premise that mathematical models are a prerequisite for scientific maturity, a stance to which the anthropologist within me is attracted. After all, we should have plenty of room for theoretical, methodological, and epistemological diversity in this field, which celebrates cultural diversity of all kinds, despite our proclivity to look down upon those who take approaches to learning in anthropology that differ from our own (Schiffer 1988, 1996, 1999).

The scientist within me, though, cannot help but sense that there is some truth to their statement. For one thing, as archaeologists we are studying the past behavior of a biological organism, and as such we are in a sense working within the biological realm of ethology, if not paleontology or paleobiology. Thus, one could argue that what is true for biology should also be true for archaeology. For another, when we rely primarily on qualitative narrative model-building, we may be and likely are occasionally building models that are in fact logically invalid, and in this one very simple sense it seems clear that a "mature science" should avoid taking such risks when possible.

Thus, I would prefer to see archaeology move more in the direction of biology. I would like to see archaeologists "put on their mathematical glasses," to quote Fagerström et al. (1996), but I do not think archaeology is ready to do so. Most of us simply do not have the educational background to construct or even evaluate formal models. For example, in a recent review of archaeological theory, Michelle Hegmon wrote (2003:229): "The mathematical and computational complexity involved in agent-based modeling and complexity theory suggests that neither will become mainstream applications in archaeology." I admit that I begrudgingly agree with Hegmon, and in a similar vein I would not be surprised if you largely glossed over all those pages with equations in this book due in part to an aversion to "mathematical complexity." We are social scientists, and most American archaeologists have been trained in the four-field approach. Unless we have sought it, we often have little training in mathematics, computer science, the physical, earth, and life sciences. Instead, we are trained primarily in largely qualitative approaches, with perhaps the single and important exception of statistical methods. I do not think it is a high priority for anthropology departments to make calculus and a course in computer programming a prerequisite for graduation in the near future, but I would like to see more students of anthropology move in this direction.

Returning to David Clarke's reason number 4 for concerning ourselves with models, a model cannot exist in the absence of a larger underlying theoretical paradigm. The presupposed underlying theory or paradigm in this study is behavioral ecology, wherein optimizing decisions are made within an environmental context—a body of theory well adapted to the study of lithic technology, which is fundamentally an exercise in problem solving. And as I have argued, the goal of the behavioral ecological approach to technology is identical to that of those studies falling under the rubric of "technological organization"—to explain technological variability as optimizing solutions governed by environmental contexts or constraints. Where they differ is at the stage of model building, specifically the use of formal models in behavioral ecology versus informal narrative models in "technological organization" research. As Clarke notes, as long as we are making models to study the same phenomena, we should build them rigorously and explicitly. I hope this study has provided at least a few examples of the utility and versatility of this approach.

## Constraints, Currencies, and Decision Variables

I have repeatedly emphasized three primary constraints on technology: mobility, raw material availability, and site function. With respect to mobility, I have suggested that the design of Folsom bifaces conforms well to the expectation of maximization of transport efficiency (amount of usable edge per unit mass). Flake tools may have been designed to maximize functional efficiency, although transport costs may have been an important secondary constraint. I have also argued that mean per capita occupation span is an excellent predictor of many aspects of technological variability. In the sites of the study sample, for example, long-term occupations, or infrequent residential mobility, equate to the accumulation of large surpluses, low degrees of tool maintenance, and increased selectivity of blanks in tool manufacture. The latter two ideas, I suggest, are not conditioned by mobility directly but instead by increased raw material availability resulting from lithic stockpiling. Therefore, there is an inextricable relationship between mobility and raw material availability that makes the assignment of causal links somewhat difficult. I have argued further that raw material availability is a function not only of natural sources of stone but also of the size of the stockpile on hand, the rate at which stone is consumed, and the rate at which a forager approaches a natural source of lithic raw material. Other constraints I discussed but did not directly examine are those imposed by flintknapping skill and function. Obviously, there are many more potential constraints on lithic technology, such as risk, social context, and raw material quality.

I have examined numerous currencies and decision variables. In chapter 5, I developed and tested a model of lithic surplus accumulation that predicted the factors governing the size of a surplus to be maintained (the decision variable) given a goal of minimization of work performed in lithic procurement (the currency). In chapter 6, I tested and expanded Kuhn's (1994) model of the design of mobile toolkits. The models predicted optimal morphological characteristics (the decision variables) of tools, cores, and bifaces, assuming a goal of maximization of tool utility per unit mass (the currency). In chapter 7, I explained changing frequencies of bifacial and core reduction debitage as a by-product of two optimization processes governing tool manufacture, use, and discard. I employed the marginal value theorem to predict at what stage in the use-lives of flake tools discard should occur (the decision variable) to maximize work performed per unit time (the currency) given the costs of tool replacement. I also modeled the relationship between raw material availability and the number of flakes produced in tool production (the decision variable) to maximize tool work efficiency (the currency). Given the immense number of potential decision-making nodes in the use-life of any given chunk of stone, this brief list is only the tip of the iceberg in terms of possible avenues of research.

## A Few Observations about Paleoindians

As I have repeatedly emphasized, this study is first about modeling, second about lithic technology, and third about Paleoindians. The models I have developed herein have general applicability, as they could be applied to any context, past, present, or future, in which stone tools are used and their assumptions are met, from the earliest toolmakers of the Oldowan to modern-day flintknappers. Their specific application to five late Pleistocene archaeological sites in the northwestern Plains provides some insight into Folsom and Goshen lifeways in that region.

### Occupation Span and Mobility

Folsom hunter-gatherers are commonly argued to have been among the most residentially mobile peoples of all time. This idea is founded on the regular occurrence in Folsom sites of nonlocal lithic raw materials that have been transported on the scale of hundreds of kilometers from their natural source areas. Kelly and Todd (1988) proposed a view of early

Paleoindian mobility wherein groups moved quickly from kill to kill to maximize encounters with bison. Short-term occupations and frequent residential mobility results in rapid movement across large areas. Although much has been written about Folsom mobility strategies (Amick 1994a,b, 1996, 2000; Blackmar 2001; Boldurian 1991; Hofman 1991, 1992, 1999a; Jodry 1999; MacDonald 1998b, 1999), prior to this study there have been no systematic attempts to monitor occupation span across a range of Folsom sites. I have provided clear evidence that there is a large range of variability in mean per capita and actual occupation spans for Folsom sites. While sites like Carter/Kerr-McGee and Agate Basin represent relatively short-term occupations, other sites, such as Barger Gulch, Locality B, Bobtail Wolf, and likely Hell Gap, represent relatively long residential stays in single locations. Krmpotich, Mill Iron, and possibly Hanson represent sites occupied for intermediate durations.

In chapter 4, I provided estimates of relative occupation span based on artifact densities. Assuming constant rates of artifact discard, I suggested that the occupation of Barger Gulch was approximately thirty times the length of that of Carter/Kerr-McGee, indicating that Barger was likely occupied for months, rather than days or weeks. In chapter 7, however, I suggested that discard rate of core reduction debitage likely increases as occupations are prolonged. This finding calls into question the accuracy of my original estimates. Because I also suggested that the rate of bifacial thinning flake discard should not respond to occupation, the density of bifacial thinning debris should provide a more accurate indicator of relative occupation span than total artifact densities. In table 8.1, I show the original and revised estimates of occupation span based on total and bifacial thinning flake densities, respectively, for all artifacts larger than 2 cm. Using only the densities of bifacial thinning flakes, the occupation of Barger Gulch is estimated instead to be approximately ten times the length of Carter/Kerr-McGee. Unless the occupation of Carter/Kerr-McGee lasted less than three days, this analysis still suggests that Folsom hunter-gatherers remained at Barger for at least one month. If so, then it is difficult to argue that Folsom residential mobility was always frequent and that the "kill-to-kill" model of residential movement always applied.

To explain lengthy stays at some sites, I proposed that snow cover and enhanced environmental patchiness with respect to fuel sources in the northwestern Plains and Rocky Mountains would have increased the cost

TABLE 8.1. Original and revised relative occupation spans for the sites of the study sample using artifacts larger than 2 cm.

| | Site | | | | |
|---|---|---|---|---|---|
| Variable | CKM | AGB | KRMP | BGB | UTM |
| Total artifacts | 40 | 532 | 584 | 1,688 | 12 |
| Bifacial thinning flakes | 13 | 177 | 99 | 197 | 0 |
| Excavation area | 34 | 171 | 74 | 51 | 25.25 |
| Total density | 1.18 | 3.11 | 7.89 | 33.10 | 0.48 |
| BFT density | 0.38 | 1.04 | 1.34 | 3.86 | 0.00 |
| Relative occupation span[†] | | | | | |
| Original estimate | 1.00 | 2.64 | 6.69 | 28.05 | 0.41 |
| Revised estimate | 1.00 | 2.72 | 3.52 | 10.17 | N/A |

*Note*: [†]Standardized to the artifact and bifacial thinning flake density of the Carter/Kerr-McGee site.

of winter residential mobility. Therefore, I argued that long-term occupations might represent winter campsites, a hypothesis that is difficult to test given the limited seasonality data available. A second, though indirect, test of this hypothesis is possible. If there is a bimodal, seasonally contingent settlement pattern, we might expect to see bimodality in the distribution of the occupation span index across a large number of sites. This test is also limited by the availability of data, but because it is not dependent on the recovery of bone, this test likely could be performed given the number of Folsom sites investigated to date. It will, however, have to wait until those data are made available.

## Site Types

Three primary site types are commonly recognized for the Folsom period: camps, kills, and quarries, although other types are occasionally described (e.g., Hofman and Ingbar 1988). Kill sites are perhaps most recognizable because of their unique character, with dense or scattered bison bone beds, often showing little butchery and frequent articulation of skeletal elements. These sites typically contain limited lithic assemblages, with relatively high frequencies of projectile points and flake tools. Camps and quarries, however, are more difficult to distinguish, even though the two terms evoke very different mental images of the nature of activities performed. Numerous undeniable quarry sites exist in Wyoming and Colorado (and around the world for that matter), the most famous of which is Spanish Diggings in eastern Wyoming (Black 2000; Coffin 1951; Miller 1991, 1996; Reher 1991). These sites, presumably used and reused over many centuries or millennia, typically exhibit shallow or deep quarry pits ringed by tailings piles and dense scatters of surface artifacts. The primary reason for the occupation of these locations is quite certainly the acquisition of lithic raw materials, but some habitation is also evident (Reher 1991).

Most Paleoindian sites that are referred to as quarries or workshops exhibit a different character. Most importantly they show no evidence of massive amounts of energy devoted to raw material procurement as seen at classic quarry sites. For example, they lack quarry pits or tailings piles. What these sites do have in common is proximity to natural sources of lithic raw material, dominance of local raw materials, and relatively high densities of artifacts. Since their discovery, the Hanson, Adair-Steadman, Lindenmeier, Bobtail Wolf, Young-Man-Chief, and Barger Gulch Folsom sites have all been referred to as quarry or workshop locations (Ingbar 1992, 1994; Root 2000; Shifrin 2000; Stanford 1999:302; White 1999; William 2000). It is not clear why many Paleoindian archaeologists are content to refer to these sites as quarries, since they differ dramatically from later quarry sites, but I suggest that we do so because they do not fit our preconceptions about what a Paleoindian campsite assemblage should look like. Folsom sites are commonly dominated by lithic raw materials from very distant sources, reinforcing the notion of super-residential mobility. When a site does not fit this pattern, it is possible to "explain away" this atypical variability by assigning a different function to the site. A quarry or workshop site should, by its very nature, be occupied and reoccupied many times, dominated by local raw materials, and exhibit high densities of artifacts. While I fundamentally agree with this view, I do not agree that only quarries will exhibit these features, or that there is any evidence whatsoever to demonstrate that the Folsom sites in question are anything more than campsites with extended occupations.

Throughout this study, I have shown that the lithic assemblage from Barger Gulch, when compared to those from Agate Basin, Carter/Kerr-McGee, and Krmpotich, is not anomalous. It is certainly extreme in many ways (high artifact density, overwhelmingly dominated by local raw materials, large quantities of surplus stone, lightly retouched tools, etc.), but it also fits almost every trend defined by the remaining three campsites. In contrast, in many analyses Upper Twin Mountain, the only kill site in my sample, breaks from the observed pattern. These observations reinforce the notion that the occupation of Upper Twin Mountain was in fact of a very different nature than the other four sites. If Barger Gulch is a quarry site, and its occupation was focused almost exclusively on the procurement of lithic raw materials, then it might be expected to deviate as well. It does not. Furthermore, use-wear studies of tools made on local raw materials at Barger Gulch and the Lake Ilo sites demonstrate that tools made locally were also used on site (Daniele 2003; Root et al. 1999), clearly indicating that activities beyond raw material acquisition and reduction were occurring. I do not mean to imply that all of the sites that have been called quarries are necessarily camps, or that Paleoindian quarry sites do not exist. I only want to make the point that by defining the problem away (i.e., treating camps with lots of local stone as quarries by default), we may reinforce our preconceptions of how the Paleoindian record should appear, but we ignore variability in Paleoindian mobility.

If I am correct that many of the Folsom sites commonly called quarries instead represent long occupations, does this mean that Paleoindians were not highly mobile? In fact, I think it says the opposite. Surely it means Folsom hunter-gatherers occasionally settled down for weeks or months, but the fact remains that stone was still being transported incredible distances by pedestrian foragers. Because a group cannot engage in residential mobility while simultaneously occupying a residential location, this implies that Folsom groups had less time available to move the hundreds of kilometers across the landscape that they apparently did. In other words, when they operated under a regime of frequent residential mobility, it must have been to an extreme degree. Whether phases of high residential mobility were characterized by frequent moves (short occupation spans), long moves between camps, or both remains to be determined.

Another observation concerning site location is in order. All of the sites I would consider candidates for long-term occupations are located near natural sources of lithic raw materials. This could be seen as another reason to argue that they represent quarry or workshop locations. However, all of the sites in the study sample are located near natural sources of lithic raw material and yet they show a significant range of variability in lithic assemblages. Proximity to raw material alone does not a quarry make. Furthermore, in long-term occupations, your raw material needs are greatest. That is, the longer you remain in one place, the greater your chances of exhausting your current supply of stone; therefore, in choosing locations to temporarily settle down, inexpensive raw material procurement would be one key variable. In fact, I hypothesize that these sites are located in areas where you find resource convergence. In particular, they are localities where reliable sources of water, food, fuel, and stone can all be procured at relatively low cost. If I am correct, this hypothesis has obvious implications for where additional extensively occupied campsites should be found.

*Raw Material Conservation*

As I discussed in depth in chapter 2, Paleoindians are commonly argued to have been lithic conservationists. Many aspects of Folsom raw material use are argued to have been responses to extreme raw material scarcity under conditions of high mobility and low availability (Ahler and Geib 2000; Amick 1995; Bamforth 2006; Boldurian and Hubinsky 1994; Hofman 1991, 1992, 1999a; Ingbar and Hofman 1999; Root et al. 1999). I agree that certain reduction strategies and tool forms were excellent solutions to raw material conservation in contexts of scarcity. Extensive tool maintenance and recycling can be viewed in the same light. I also agree that these behaviors were commonly employed, but they were employed only under certain conditions.

I have modeled various ways in which there are advantages to being "wasteful." Common behaviors that did occur and could be construed as examples of wastefulness include the use of amorphous core reduction, the abandonment of usable raw material, the discard of tools long before they exhausted their full potential use-lives, and the failure to use many potential tool flakes. At Barger Gulch, for example, I have estimated that we have recovered slightly over 7 kg of surplus lithic raw material. Because we have excavated only a small portion of the site, the total surplus at Barger is likely much larger. If we have a 10 percent sample (likely an overestimate), 70 kg, or more than 150 lbs, of raw material may have been brought into the site only to be abandoned upon departure. It is difficult to construe such behavior as conservation in any way. Clearly, under some conditions, Folsom hunter-gatherers showed little concern for raw material conservation, but I have shown that they may have done so for good reasons. Accumulating surpluses prevents lithic shortfalls and reduces the costs of tool replacement. When tools can be replaced at low costs, I argue, as have others (Kuhn 1989, 1995; Parry and Kelly 1987), average tool efficiency can be enhanced. In this light, behaviors that could be construed as wasteful are nothing more than optimal solutions to specific environmental contexts and are not wasteful at all.

*A Sidebar: A Partial Solution to the Hanson Problem*

I began chapter 3 by discussing alternative interpretations of mobility, reoccupation, and raw material use at the Hanson site in the Bighorn Basin of Wyoming. The primary area of disagreement was whether Phosphoria Formation chert, located approximately 20 km from the site, was transported into the site from the previous residential camp (a nonlocal raw material by my usage), or was acquired during the occupation of the site (a local raw material). Based on patterns of raw material representation, Ingbar (1992, 1994) argued that Phosphoria chert was brought into the site at the time of initial occupation. Frison and Bradley (1980), noting the same pattern, suggest that Phosphoria chert was simply preferred for the production of projectile points and endscrapers, and like Amick (1994a) argue that it was acquired during the occupation of the site. Similar differences exist in the interpretation of site occupation history. Ingbar (1992) argues that the Hanson assemblage is the product of numerous short-term occupation events, while Frison and Bradley (1980) and Amick (1994a) suggest that relatively few extended occupations are present. Stratigraphic evidence suggests that minimally two occupations are present (Frison and Bradley 1980).

In chapter 3, I was unable to resolve these differences. Using data from the study sample and published sources, I plotted the ratios of local:nonlocal raw material and debitage:nonlocal tools for eleven Folsom and Goshen components (fig. 3.9). The position of the Hanson site on that graph depended on whether Phosphoria chert was considered local or nonlocal. Phosphoria chert not only falls on the 20 km cutoff distance I use to distinguish local from nonlocal raw material, but it also fits the trend

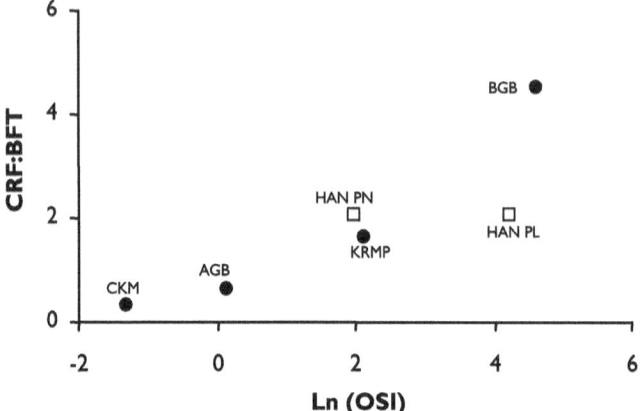

FIGURE 8.1. Ln (occupation span index) vs. the ratio of core reduction to bifacial thinning flakes for the four campsites of the study sample showing the position of the Hanson site for two calculated OSIs assuming Phosphoria chert is local (HAN PL) or nonlocal (HAN PN).

defined by the remaining ten sites in both cases—if it is considered local or nonlocal. If Phosphoria is treated as a local raw material, the mean per capita occupation span of Hanson falls with Barger Gulch and Bobtail Wolf, suggesting long-term occupation. If it is treated as nonlocal, Hanson falls close to Mill Iron and Krmpotich, suggesting intermediate length occupation. Using debitage ratios may provide a way out of this morass.

In chapter 7, I showed that as occupations are prolonged, sites become increasingly dominated by core reduction debitage. Because Ingbar (1992:table 5.2) provides counts of bifacial thinning and core reduction debitage for the Hanson site, it is possible to ask under what condition (Phosphoria local or nonlocal) the Hanson assemblage best fits the trend defined by the four campsites. Ingbar (1992:table 5.2) records 1,413 bifacial thinning flakes and 2,936 core reduction flakes for the assemblage, resulting in a core reduction to bifacial thinning flake ratio of 2.08. Comparing these values to those of the four campsites (table 8.1) provides a solution.[1] Figure 8.1 shows that when Phosphoria chert is treated as a nonlocal raw material, Hanson almost perfectly fits the pattern defined by the remaining four campsites. If Phosphoria is classified as local, then Hanson breaks from the trend considerably. This would support the hypothesis that Phosphoria chert should be considered a nonlocal if not predominantly nonlocal raw material.

Therefore, Hanson appears to have a mean per capita occupation span of intermediate length roughly comparable to that of the Krmpotich site. Unfortunately, it is not possible to estimate the actual or cumulative occupation span of Hanson, since it is known to contain at least two occupations and comparable artifact densities cannot be calculated. In the simplest cases, Hanson could, for example, be composed of two intermediate-length occupations or one long- and one short-term occupation, but with the data currently available it is not possible to evaluate the site further. What is clear is that Phosphoria chert was likely transported into the site in the form of bifaces and finished tools as Ingbar (1992) has suggested, and little if any Phosphoria was acquired during the occupation of the site. Obviously, I would disagree with Ingbar's (1992, 1994) contention that Hanson was a quarry or workshop site, and based on my finding of no reoccupation in the study sample, I suspect that only two or very few occupations are present, a hypothesis that will have to wait until additional data are available for further evaluation.

*Folsom Variability in Context*

In a 1993 paper, David Meltzer asked, "Is there a Clovis adaptation?" He concluded that Clovis adaptations likely varied by region, and that we cannot uncritically take a single model of Clovis subsistence and apply it universally to the entirety of North America. We could ask the same question of Folsom technology. After all, the Folsom complex spans a huge area and many biogeographic zones. In fact, it is becoming rather commonplace to question whether Folsom lifestyles were consistent from Texas to Alberta and from Utah to the Midwestern prairie states, with many authors noting or suggesting regional variability (Amick 1994a, 1996; LeTourneau 1998; MacDonald 1999). In this study, I have explored technological variability in five sites from a large area that I have referred to as the "Northwestern Plains and Rocky Mountains," a simple unifying term, but one that masks environmental variability.

The five sites span an area of more than 200,000 km$^2$, approximately 1,500 m in elevation, and considerable environmental variability, ranging from the sand dunes of southwestern Wyoming to the shortgrass prairies of eastern Wyoming to the semi-arid intermontane basin of Middle Park, Colorado, where big sagebrush dominates the valley bottom. Surely, we might expect to see a natural environmental signal affecting the assemblages, and we do. I have shown that there is considerable variability among the assemblages of the study sample, and I have primarily discussed variation in lithic assemblage with respect to three environmental constraints—mobility, raw material availability, and site function. As all of these constraints can be expected to vary with respect to region and site, it should come as no surprise that assemblages are not entirely redundant in character. It is important to note, however, that the variation I have documented can be explained by the operation of very simple rules under differing contexts.

This observation raises an important point. When we identify differences in Folsom assemblages or adaptations between the Northern Plains and Southern Plains (MacDonald 1999) or the Llano Estacado and the large and small valleys of the Basin and Range (Amick 1994a, 1996, 2000), is it because different rules governed the use of lithic raw materials in these areas or because people, using the same rules, operated under different constraints? I expect that the truth lies closer to the latter.

Confirming or falsifying this hypothesis will require further research. One simple test could involve the analysis of lithic assemblages from sites of similar raw material contexts (close proximity to natural sources of raw material) in other regions. For example, if the rules governing Southern Plains Folsom lithic technology were identical to those of the Northern Plains and Rocky Mountains, then the lithic assemblages at sites, such as Adair-Steadman (Tunnell 1977), located near a source of Edwards chert, should conform to the trends identified in the study sample. One problem with adequately addressing this issue is that many Folsom sites are located in regions wherein no high-quality lithic raw materials are available within 20 or even 50 km distance. The various models I have developed are in part dependent on this constraint, and as many others and I have suggested, raw material availability is one primary constraint shaping technological strategies. As such, additional work is needed to allow direct comparison of sites in areas rich and poor in raw material by the development of rigorous archaeological measures of raw material availability. Furthermore, it may also be necessary to develop measures of mean per capita occupation span that are not dependent on the use of lithic raw material data since many Folsom sites contain no local raw materials.

## A Final Observation

I end this study with one final observation concerning the approach I have used. As I have written many times, the fundamental goals of the models that I have developed and tested are to define optimal technological strategies for any given environmental context. I have looked at numerous aspects of technological variability and have done so without once applying the concepts of curation, expediency, versatility, flexibility, reliability, or maintainability. While these conceptual categories have inspired many years of research, people have noted that these concepts are very difficult to apply because they are nebulously defined and difficult to operationalize (e.g., Bamforth 1986; Hayden et al. 1996; Nash 1996; Shott 1996). Instead, I have relied exclusively on much simpler concepts, such as transport, use-life, consumption rate, discard rate, maintenance, selectivity, and so on. By doing so, I was able to perform relatively straightforward tests of the models developed and capture significant intersite variability. I suggest that it is time that we step back from the body of work that has come to be known as "technological organization," a paradigm that has fundamentally changed for the better the ways that archaeologists view stone tool technology, and instead build simple mathematical models from very simple component variables. Before this study, others had begun to do so (Brantingham and Kuhn 2001; Kuhn 1994; Metcalfe and Barlow 1992), and if we continue in this framework, I believe our understanding of the factors governing variability in stone tool technology throughout space and time will grow tremendously.

# Appendix
## Site Occupancy and Camp Area

IN THIS APPENDIX, I discuss the relationship between the number of site occupants and camp area in order to justify the simplified model of site area used in chapter 3. It is important to note that site area is more closely tied to the number of households than the number of occupants. Obviously, these two variables should be highly correlated. For the following equations, the term $p$ representing the number of site occupants could be divided by a constant representing the number of individuals per household or could simply be substituted with a variable representing the number of households. This omission is of little consequence since it would have no impact on the form of any of the equations.

I begin with a circular camp model based loosely on Yellen's (1977) "ring model." Consider a circular camp with a circumference ($c$) governed by the number of occupants ($p$)

$$c = pl \tag{A.1}$$

where $l$ is the per capita distance requirement for occupants arranged linearly. For any circle, the relationship between radius ($r$) and circumference is:

$$c = 2\pi r$$
$$r = \frac{c}{2\pi} \tag{A.2}$$

Therefore, the camp radius can be expressed as a function of the number of site occupants:

$$r = \frac{pl}{2\pi} \tag{A.3}$$

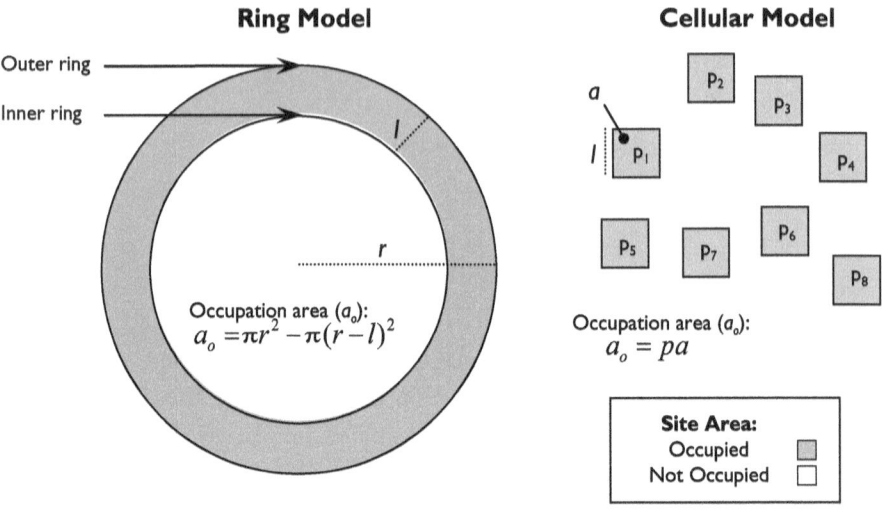

FIGURE A.1.

Using this equation, it is also possible to express site area ($s$) as a function of the number of site occupants:

$$s = \pi r^2$$

$$s = \pi \left(\frac{pl}{2\pi}\right)^2$$

$$s = \frac{p^2 l^2}{4\pi} \quad (A.4)$$

As discussed in chapter 3, for circular camps camp area is expected to increase as a function of the per capita area requirements ($l^2$) and the square of the number of site occupants. Assuming that archaeological excavations focus on areas occupied and that the ring of occupation maintains a constant thickness of $l$ (fig. A.1) governed by the size of households and related working areas, area occupied ($s_o$) is instead calculated as the difference in area between two concentric circles:

$$s_o = \pi r^2 - \pi(r-l)^2$$

$$s_o = \pi(2rl - l^2) \quad (A.5)$$

Combining equations A.3 and A.5 gives the relationship between the number of site occupants and the area occupied:

$$s_o = \pi \left( 2 \left( \frac{pl}{2\pi} \right) l - l^2 \right)$$

$$s_o = pl^2 - \pi l^2 \qquad (A.6)$$

Therefore, area occupied should increase as a linear function of the number of site occupants. Though this equation should accurately portray the relationship between area occupied and site population size for circular camps, it is problematic in that it has a negative y-intercept. In other words, a camp with zero occupants is expected to have a negative area occupied. Furthermore, if small values are inserted for the variable $p$ ($p < \pi$), negative camp areas are also predicted. The problem stems from the fact that one or two people or households cannot camp in circles, and if site radius exceeds the thickness of the occupied zone, a negative area is expected.

Therefore, a cellular model of habitation (fig. A.1) may be more desirable, where the "cells" of occupation are arranged in any form, whether random, linear, circular, elliptical, rectangular, and so on. In this case, area occupied ($l_c$) should be a function of the number of site occupants ($p$), and the per capita area requirement ($a$):

$$a_o = pa \qquad (A.7)$$

If archaeologists excavate only or primarily occupied areas, then artifact densities ($d/s_o$) should be independent of the number of site occupants because the number of site occupants cancels out from the equation:

$$\frac{d}{s_o} = \frac{prt}{pa}$$

$$\frac{d}{s_o} = \frac{rt}{a} \qquad (A.8)$$

where $r$ is the per capita discard rate and $t$ is the occupation span. Thus, one would be justified in using artifact densities as a measure of cumulative occupation span if (1) large empty areas of sites are not excavated; (2) discard rates ($r$) are assumed to be constant; and (3) per capita space requirements ($a$) are assumed to be constant.

# Notes

## Chapter 1

1. The applicability of the zero-one rule hinges on the nature of the gain function describing processing time and load utility. If the gain function is linear, then a zero-one rule applies. If the function is curvilinear down, the zero-one rule no longer applies, and partially processed loads become optimal.

2. In chapter 6, it is shown that the second prediction is incorrect. The optimal size for tool blanks is 1.5 to 2.0 the minimum usable portion, depending on the way utility is measured. This may seem odd since I have argued that logically speaking, formal models have predictions that must follow from their assumptions. Of course, this is true only if no errors are made in calculations or in manipulations of equations.

3. As is the case for much of oral folklore, I do not know the original source of this joke. It was told to me by John Moore, professor of chemistry at the University of Wisconsin–Madison in the early 1990s.

## Chapter 3

1. Following Deboer (1974), I have generalized David's (1972:142) equation, which was originally written to calculate the number of pots discarded after one hundred years, and I have modified the notation so as to be consistent with Schiffer's (1975a,b) discard equation.

2. It is a simple matter to show that this equation is a modified version of Schiffer's (1975a,b) discard equation (equation 3.2 in this chap.). If the term $S$ in Schiffer's equation (the number of artifacts in systemic context) is replaced with $p \bullet a$ representing the number of site occupants multiplied by the number of artifacts maintained per person, the equation becomes: $d_t = p \dfrac{a}{L} t$. Therefore, the number of artifacts discarded is equal to the product of the number of site occupants ($p$), discard rate ($a/L$), and occupation span.

3. There is one small caveat to this statement. If two types of artifacts have discard rates that change proportionally, artifact ratios will remain constant. For this approach to be used, the discard rate of at least one of the artifact types must change with time and the discard rate of the other artifact type must not change proportionally.

4. I use a nonparametric correlation coefficient because Pearson's r is very sensitive to outliers. If one point in a Monte Carlo run was a severe outlier, a strong correlation would result. In contrast, Spearman's ρ is completely insensitive to outlying values.

## Chapter 5

1. For additional realism, equation 5.7 could be multiplied by a constant representing the probability of a lithic deficit if no surplus is maintained. Presumably, the variable $p$ should never be equal to 1 because even if no surplus is maintained, there is a decent chance that on any given day a deficit will not occur owing to a low rate of lithic consumption. Therefore, the maximum value of $p$ in actuality should be somewhat less than one. In general, this constant should be governed by the degree of dispersion in both rates of supply and demand.

2. It is fairly straightforward to show that artifact proportions are independent of the number of site occupants, that is, they are per capita measures. If every individual maintains a surplus, the size of the surplus (the numerator of equation 5.12) would be multiplied by a variable representing the number of site occupants. If every individual is also responsible for consumed raw material, the denominator of equation 5.12 would be multiplied by the same constant. Thus, the number of individuals cancels out.

3. The following artifact types were considered to represent "surplus raw material": types 1a (large cores, 1 striking platform), 1b (horse hoof cores), 1c (large cores, >1 striking platform), 2a (large flake scrapers with retouch), 2b (spokeshaves), 3a (tula adzes), 3c (non-tula adzes). Included within "consumed raw materials" were types 1d (micro-cores, 1 striking platform), 1e (micro-cores, >1 striking platforms), 3b (tula adze slugs), 3d (non-tula adze slugs), 3e (tula micro-adzes), 3f (tula micro-adze slugs), 3g (non-tula micro-adzes), 3h (non-tula micro-adze slugs), 3i (small endscrapers), 4a (lunates), 4b (Bondi points), 4c (backed blades of irregular shape), 9 (retouched frag. and utilized flakes), and debitage.

4. Here is where formality meets informality. I am stumped by the problem of coming up with a formal definition of the distinction between "surplus" and "consumed" raw material. The 5-cm size cutoff was really one of convenience because sample sizes for the study sample become extremely small if a larger size cutoff is used.

5. The number of forays was decreased because adding multiple sources to the simulation increased computing time considerably. The difference in the standard error of the estimated mean embedded procurement costs averaged for 100 and 1,000 forays should not be substantial.

6. As I have structured the simulation, it is difficult to determine exactly how the number and spatial arrangement of raw material sources affects the $d$:$e$ ratio. Because sources are randomly positioned, for example, the number of sources and the distance to the closest source (and mean distance to all sources) tend to be somewhat correlated. One solution to this problem would be to hold number of sources constant while varying the distance to the closest source. It would also be advantageous to vary

the position of sources to examine the effect of changes in "radial coverage." For example, how is the *d:e* ratio affected by sources that cluster on one side of a site versus sources that are distributed in all directions outward from a site?

7. This estimate is based on the cores of the study sample, which range in mass from 4 to 730 g. This gives at least a rough estimate of the maximum size of lithic nodules typically transported to sites for reduction.

## Chapter 6

1. This analysis excluded artifacts that could not be confidently assigned to the "Yes" or "No" categories for the variable "Bifacial Thinning Flake."

2. The thickness:length and thickness:width ratios were assessed only for flakes on which reliable width and length measurements, respectively, could be taken (Surovell 2003a:340–353).

## Chapter 8

1. This analysis is dependent on the assumption of sample equivalence. Ingbar's (1992) sample included all items that were mapped in place from the northwest block of Area 2 of the site (Ingbar, personal communication, 2001). My analysis includes all artifacts mapped in place and items larger than 1.5 or 2 cm from screened material varying by site (Surovell 2003a:340–353). Although there may be a slight discrepancy in artifact selection since Ingbar did not include any artifacts from screens, I believe the samples to be roughly equivalent.

# References

Agogino, G. A., and W. D. Frankfurter
1959 The Brewster site: An Agate-Basin Folsom multiple component site in eastern Wyoming. *The Masterkey* 33 (1): 102–107.

Ahler, S. A., and P. R. Geib
2000 Why flute: Folsom point design and adaptation. *Journal of Archaeological Science* 27:799–820.

Alcock, J.
2001 *The Triumph of Sociobiology.* Oxford: Oxford University Press.

Aldenderfer, M. S.
1981 Creating assemblages by computer simulation: The development and uses of ABSIM. In *Simulation in Archaeology*, ed. J. A. Sabloff, 67–117. Albuquerque: University of New Mexico Press.

Alexander, R. M.
1996 *Optima for Animals.* Princeton, NJ: Princeton University Press.

Allen, E., B. Beckwith, D. Culver, M. Dincan, S. Gould, R. Hubbard, H. Inouye, A. Leeds, R. Lewontin, C. Madansky, L. Miller, R. Pyeritz, M. Rosenthal, and H. Schreier
1975 Against "Sociobiology." *New York Review of Books*, 13 November 1975, 43–44.

Amick, D. A.
2000 Regional approaches with unbounded systems: The record of Folsom land use in New Mexico and west Texas. In *The Archaeology of Regional Interaction: Religion, Warfare, and Exchange across the American Southwest and Beyond*, ed. M. Hegmon, 119–147. Boulder: University Press of Colorado.

Amick, D. S.
1994a Folsom diet breadth and land use in the American Southwest. PhD diss., University of New Mexico.
1994b Technological organization and the structure of inference in lithic analysis: An examination of Folsom hunting behavior in the American Southwest. In *The Organization of North American Chipped Stone Technologies*, ed. P. J. Carr, 9–34. Ann Arbor, MI: International Monographs in Prehistory.
1994c Edwards chert use by Folsom hunters in New Mexico. *Current Research in the Pleistocene* 11:59–61.

Amick, D. S.
1995  Patterns of technological variation among Folsom and Midland projectile points in the American Southwest. *Plains Anthropologist* 10:23–28.
1996  Regional patterns of Folsom mobility and land use in the American Southwest. *World Archaeology* 27:411–426.

Amick, D. S., and R. P. Mauldin, eds.
1989  *Experiments in Lithic Technology*. Oxford: British Archaeological Reports.

Amick, D. S., R. P. Mauldin, and S. A. Tomka
1988  An evaluation of debitage produced by experimental bifacial core reduction of a Georgetown chert nodule. *Lithic Technology* 17:26–36.

Ammerman, A. J., and M. W. Feldman
1974  On the "making" of an assemblage of stone tools. *American Antiquity* 39:611–616.

Andrefsky, W., Jr.
1991  Inferring trends in prehistoric settlement behavior from lithic production technology in the Southern Plains. *North American Archaeology* 12 (2): 129–144.
1994  Raw material availability and the organization of technology. *American Antiquity* 59:21–34.
1998  *Lithics: Macroscopic Approaches to Analysis*. Cambridge: Cambridge University Press.
2001  Emerging directions in debitage analysis. In *Lithic Debitage: Context, Form, Meaning*, ed. W. Andrefsky Jr., 2–14. Salt Lake City: University of Utah Press.

Aveleyra Arroyo de Anada, L.
1961  El primer hallazgo Folsom en territorio Mexicano y su relación con el complejo de puntas acanaldas en Norte América. In *Homenaje a Pablo Martínez del Río en el XXV Aniversario de la Edición de Los Orígenes Americanos*, 31–48. Mexico City: Instituto Nacional de Antropología e Historia.

Ballenger, J.
1999  Late Paleoindian land use in the Oklahoma Panhandle: Goff Creek and Nall Playa. *Plains Anthropologist* 44:189–207.

Bamforth, D. B.
1985  The technological organization of Paleo-Indian small-group bison hunting on the Llano Estacado. *Plains Anthropologist* 30:243–258.
1986  Technological efficiency and tool curation. *American Antiquity* 51:38–50.
1990a  Settlement, raw material, and lithic procurement in the Central Mojave Desert. *Journal of Anthropological Archaeology* 9:70–104.
1990b  Technological organization and hunter-gatherer land use: A California example. *American Antiquity* 56:216–234.
2002a  Evidence and metaphor in evolutionary archaeology. *American Antiquity* 67:435–452.
2002b  High-tech foragers? Folsom and later Paleoindian technology on the Great Plains. *Journal of World Prehistory* 16:55–98.

Bamforth, D. B.
2006  The Windy Ridge quartzite quarry: Hunter-gatherer mining and hunter-gatherer land use on the North American continental divide. *World Archaeology* 38:511–527.

Bamforth, D. B., and M. S. Becker
2000  Core/biface ratios, mobility, refitting, and artifact use-lives: A Paleoindian example. *Plains Anthropologist* 45:273–290.

Bamforth, D. B., and P. Bleed
1997  Technology, flaked stone technology, and risk. In *Rediscovering Darwin: Evolutionary Theory in Archeological Explanation*, ed. C. M. Barton and G. A. Clark, 109–139. Arlington, VA: American Anthropological Association.

Bass, W. M.
1970  *Excavations of a Paleo-indian Site at Agate Basin, Wyoming*. National Geographic Society Research Reports: 1961–1962 Projects.

Beaton, J. M.
1991  Colonizing continents: Some problems from Australia and the Americas. In *The First Americans: Search and Research*, ed. T. D. Dillehay and D. J. Meltzer, 209–230. Boca Raton, FL: CRC Press.

Beck, C., A. K. Taylor, G. T. Jones, C. M. Fadem, C. R. Cook, and S. A. Millward
2002  Rocks are heavy: Transport costs and Paleoarchaic quarry behavior in the Great Basin. *Journal of Anthropological Archaeology* 21 (4): 481–507.

Belovsky, G. E.
1987  Hunter-gatherer foraging: A linear programming approach. *Journal of Anthropological Archaeology* 6:29–76.

Bement, L. C.
1997  The Cooper site: A stratified Folsom bison kill in Oklahoma. *Plains Anthropologist* 42 (159): 85–99.
1999  *Bison Hunting at the Cooper Site: Where Lightning Bolts Drew Thundering Herds*. Norman: University of Oklahoma Press.

Billeck, W. T.
1998  Fluted point distribution in the Loess Hills of southwestern Iowa. *Plains Anthropologist* 43:401–409.

Binford, L. R.
1973  Interassemblage variability—the Mousterian and the "functional" argument. In *The Explanation of Culture Change: Models in Prehistory*, ed. C. Renfrew, 227–254. London: Duckworth.
1977  Forty-seven trips: A case study in the character of archaeological formation processes. In *Stone Tools as Cultural Markers*, ed. R.V.S. Wright, 24–36. Canberra: Australian Institute of Aboriginal Studies.
1979  Organization and formation processes: Looking at curated technologies. *Journal of Anthropological Research* 35 (3): 255–273.
1980  Willow smoke and dogs' tails: Hunter-gatherer settlement systems and archaeological site formation. *American Antiquity* 45:4–20.

Binford, L. R.
1983    *In Pursuit of the Past: Decoding the Archaeological Record.* New York: Thames and Hudson.
1991    When the going gets tough: Nunamiut local groups, camping patterns, and economic organisation. In *Ethnoarchaeological Approaches to Mobile Campsites*, ed. C. S. Gamble and W. A. Boismier, 25–138. Ann Arbor, MI: International Monographs in Prehistory.
2001    *Constructing Frames of Reference: An Analytical Method for Archaeological Theory Building Using Ethnographic Data Sets.* Berkeley and Los Angeles: University of California Press.

Binford, L. R., and J. F. O'Connell
1986    An Alyawara day: Making men's knives and beyond. *American Antiquity* 51:547–562.
1989 [1984]   An Alyawara day: The stone quarry. In *Debating Archaeology*, ed. L. R. Binford, 406–432. San Diego: Academic Press.

Bird, D. W., and J. F. O'Connell
2006    Behavioral ecology and archaeology. *Journal of Archaeological Research* 14:143–188.

Black, K. D.
2000    Lithic sources in the Rocky Mountains of Colorado. In *Intermountain Archaeology*, ed. D. B. Madsen and M. D. Metcalf, 132–147. Salt Lake City: University of Utah Press.

Blackmar, J. M.
2001    Regional variability in Clovis, Folsom, and Cody land use. *Plains Anthropologist* 46:65–94.

Bleed, P.
1986    The optimal design of hunting weapons: Maintainability or reliability. *American Antiquity* 51:737–747.

Bliss, W. L.
1939    Early man in western and northwestern Canada. *Science* 89:365–366.

Boldurian, A. T.
1990    *Lithic Technology at the Mitchell Locality of Blackwater Draw: A Stratified Folsom Site in Eastern New Mexico.* Memoir 24. Lincoln, NE: Plains Anthropological Society.
1991    Folsom mobility and organization of lithic technology: A view from Blackwater Draw, New Mexico. *Plains Anthropologist* 36:281–295.

Boldurian, A. T., G. A. Agogino, P. H. Shelley, and M. Slaughter
1987    Folsom biface manufacture, retooling, and site function at the Mitchell locality of Blackwater Draw. *Plains Anthropologist* 32:299–311.

Boldurian, A. T., and J. L. Cotter
1999    *Clovis Revisited: New Perspectives on Paleoindian Adaptations from Blackwater Draw, New Mexico.* Philadelphia: The University Museum, University of Pennsylvania.

Boldurian, A. T., and S. M. Hubinsky
1994    Preforms in Folsom lithic technology: A view from Blackwater Draw, New Mexico. *Plains Anthropologist* 39:445–464.

Boone, J. L., and E. A. Smith
1998    Is it evolution yet? A critique of evolutionary archaeology. *Current Anthropology* 39:S141–S173.

Bousman, C. B.
1993    Hunter-gatherer adaptations, economic risk, and tool design. *Lithic Technology* 18:59–86.

Bowler, J. M.
1976    Recent developments in reconstructing Quaternary environments in Australia. In *The Origin of the Australians*, ed. R. L. Kirk and A. G. Thorne, 55–77. Canberra: Australian Institute of Aboriginal Studies.

Boyd, R., and P. J. Richerson
1985    *Culture and the Evolutionary Process.* Chicago: University of Chicago Press.

Bradley, B. A.
1982    Flaked stone technology and typology. In *The Agate Basin Site: A Record of the Paleoindian Occupation of the Northwestern High Plains*, ed. G. C. Frison and D. J. Stanford, 181–208. New York: Academic Press.

Bradley, B. A., and G. C. Frison
1996    Flaked-stone and worked-bone artifacts from the Mill Iron site. In *The Mill Iron Site*, ed. G. C. Frison, 71–86. Albuquerque: University of New Mexico Press.

Brantingham, P. J., and S. L. Kuhn
2001    Constraints on Levallois core technology: A mathematical model. *Journal of Archaeological Science* 28:747–761.

Bright, J., A. Ugan, and L. Hunsaker
2002    The effect of handling time on subsistence technology. *World Archaeology* 34:164–181.

Broecker, W. C., D. M. Peteet, and D. Rind
1985    Does the ocean-atmosphere have more than one stable mode of operation? *Nature* 315:21–26.

Broecker, W. S.
1994    Massive iceberg discharges as triggers for global climate change. *Nature* 372:421–424.

Broecker, W. S., M. Andree, W. Wolfli, H. Oeschger, G. Bonani, J. Kennett, and K. Peteet
1988    The chronology of the last deglaciation: Implications to the cause of the Younger Dryas event. *Paleooceanography* 3:1–19.

Broecker, W. S., J. P. Kennett, B. P. Flower, J. T. Teller, S. Trumbore, G. Bonani, and W. Wolfli
1989    Routing of meltwater from the Laurentide ice sheet during the Younger Dryas cold episode. *Nature* 341:318–321.

Broughton, J. M., and J. F. O'Connell
1999    On evolutionary ecology, selectionist archaeology, and behavioral archaeology. *American Antiquity* 64:153–165.
Buchanan, B.
2002    Folsom lithic procurement, tool use, and replacement at the Lake Theo Site. *Plains Anthropologist* 47:121–146.
Butler, R., and R. J. Fitzwater
1965    A further note on the Clovis site at Big Camas Prairie, southcentral Idaho. *Tebiwa* 8 (1): 38–40.
Byerly, R. M., J. R. Cooper, D. J. Meltzer, M. E. Hill, and J. M. LaBelle
2005    On Bonfire shelter (Texas) as a Paleoindian bison jump: An assessment using GIS and zooarchaeology. *American Antiquity* 70:595–629.
Byers, D. A.
2001    The Hell Gap Site Locality II Agate Basin faunal assemblage: A study of Paleoindian raunal resource exploitation. MA thesis, University of Wyoming.
Byers, D. A., and A. Ugan
2005    Should we expect large game specialization in the late Pleistocene? An optimal foraging perspective on early Paleoindian prey choice. *Journal of Archaeological Science* 32:1624–1640.
Cannon, M. D., and D. J. Meltzer
2004    Early Paleoindian foraging: Examining the faunal evidence for megafaunal specialization and regional variability in prey choice. *Quaternary Science Reviews* 23:1955–1987.
Carpenter, L. H., R. B. Gill, D. J. Freddy, and L. E. Sanders
1979    *Distribution and Movements of Mule Deer in Middle Park, Colorado. Colorado Division of Wildlife*. Special Report No. 46. Fort Collins.
Casti, J. L.
1997    *Would-be Worlds: How Simulation Is Changing the Frontiers of Science*. New York: John Wiley and Sons.
Charnov, E. L.
1976a    Optimal foraging: Attack strategy of a mantid. *American Naturalist* 110:141–151.
1976b    Optimal foraging, the marginal value theorem. *Theoretical Population Biology* 9:129–136.
Clarke, D. L., ed.
1972a    *Models in Archaeology*. London: Methuen.
Clarke, D. L.
1972b    Models and paradigms in contemporary archaeology. In *Models in Archaeology*, ed. D. L. Clarke, 3–60. London: Methuen.
1976    Mesolithic Europe: The economic basis. In *Problems in Economic and Social Archaeology*, ed. G.D.G. Sieveking, I. H. Longworth, and K. E. Wilson, 449–481. London: Duckworth.

Coffin, R. G.
1951 Sources and origin of northern Colorado artifact materials. *Southwestern Lore* 17 (1): 2–6.

Collins, M. B.
1999 *Clovis Blade Technology*. Austin: University of Texas Press.

Cowan, F. L.
1999 Making sense of flake scatters: Lithic technological strategies and mobility. *American Antiquity* 64:593–607.

Crabtree, D. E.
1966 A stoneworker's approach to analyzing and replicating the Lindenmeier Folsom. *Tebiwa* 9:3–39.

Craig, C.
1983 Lithic source analysis and interpretation in northeastern Wyoming. MA thesis, University of Wyoming.

Daniele, J. R.
2003 The Barger Gulch Locality B formal tool assemblage: A use-wear analysis. MA thesis, University of Wyoming.

David, N.
1972 On the life span of pottery, type frequencies, and archaeological inference. *American Antiquity* 37:141–142.

Davis, L. B., and S. T. Greiser
1992 Indian Creek Paleoindians: Early occupation of the Elkhorn Mountains' southeast flank, west-central Montana. In *Ice Age Hunters of the Rockies*, ed. D. J. Stanford and J. S. Day, 225–283. Niwot: Denver Museum of Natural History and University Press of Colorado.

Dawkins, R.
1989 *The Selfish Gene*. Oxford: Oxford University Press.

de Barros, P.L.F.
1982 The effects of variable site occupation span on the results of frequency seriation. *American Antiquity* 47:291–315.

Deboer, W. R.
1974 Ceramic longevity and archaeological interpretation. *American Antiquity* 39:335–343.

Dibble, D. S., and D. Lorrain
1968 *Bonfire Shelter: A Stratified Bison Kill Site, Val Verde County, Texas*. Miscellaneous Papers No. 4, Texas Memorial Museum, University of Texas, Austin.

Dillehay, T. D.
2000 *The Settlement of the Americas: A New Prehistory*. New York: Basic Books.

Dixon, E. J.
1999 *Bones, Boats, and Bison: Archeology and the First Colonization of Western North America*. Albuquerque: University of New Mexico Press.

Donohue, J., and F. Sellet
2001   The chronology of the Goshen bone bed at the Jim Pitts site. *Current Research in the Pleistocene* 18:20–21.

Ebell, B.
1970   A Folsom site in southeastern Saskatchewan. *Saskatchewan Archaeology Newsletter* 30:17–18.

Eerkens, J. W.
1998   Reliable and maintainable technologies: Artifact standardization and the Early to Later Mesolithic transition in northern England. *Lithic Technology* 23:42–53.

Emerson, A. M.
2000   Analysis of vertebrate remains. In *The Archaeology of the Bobtail Wolf Site*, ed. M. J. Root, 309–345. Pullman: Washington State University Press.

Epstein, J.
1969   *The San Isidro Site: An Early Man Campsite in Nuevo Leon, Mexico.* Anthropology Series, Number 7, Department of Anthropology, University of Texas at Austin.

Epstein, J. F.
1961   The San Isidro and Puntita Negra sites: Evidence of early man horizons in Nuevo Leon, Mexico. In *Homenaje a Pablo Martínez del Río en el XXV Aniversario de la Edición de Los Orígenes Americanos*, 71–74. Mexico City: Instituto Nacional de Antropología e Historia.

Ewers, J. C.
1955   *The Horse in Blackfoot Indian Culture.* Washington DC: Smithsonian Institution Press.

Fagerström, T., P. Jagers, P. Schuster, and E. Szathmary
1996   Biologists put on mathematical glasses. *Science* 274:2039–2040.

Faught, M. K., and A. K. Freeman
1998   Paleoindian complexes of the terminal Wisconsin and early Holocene. In *Paleoindian and Archaic Sites in Arizona*, ed. J. B. Mabry, 33–52. Phoenix: State Historic Preservation Office, Arizona State Parks.

Ferring, C. R.
2001   *The Archaeology and Paleoecology of the Aubrey Clovis Site (41DN479), Denton County, Texas.* Denton, Center for Environmental Archaeology, Department of Geography, University of North Texas.

Fiedel, S. J.
1999   Older than we thought: Implications of corrected dates for Paleoindians. *American Antiquity* 64:95–115.

Forbis, R. G., and J. D. Sperry
1952   An early man site in Montana. *American Antiquity* 18:127–133.

Fosha, M., and F. Sellet
2000   The Ghost site, A Folsom/Goshen locality in South Dakota. *Current Research in the Pleistocene* 2000:26–27.

Francis, J., and M. L. Larson
1996    Chipped-stone raw material from the Mill Iron site. In *The Mill Iron Site*, ed. G. C. Frison, 87–100. Albuquerque: University of New Mexico Press.

Frison, G. C.
1977    The Paleo-Indian in the Powder River Basin. Paper presented at the 35th Plains Conference, Lincoln, Nebraska.
1978    *Prehistoric Hunters of the High Plains*. 1st ed. New York: Academic Press.
1982a    Raw stone flaking material sources. In Frison and Stanford, *The Agate Basin Site*, 173–178. New York: Academic Press.
1982b    Folsom components. In Frison and Stanford, *The Agate Basin Site*, 37–76. New York: Academic Press.
1982c    Radiocarbon dates. In Frison and Stanford, *The Agate Basin Site*, 178–180. New York: Academic Press.
1982d    Bison dentition studies. In Frison and Stanford, *The Agate Basin Site*, 240–260. New York: Academic Press.
1984    The Carter/Kerr-McGee Paleoindian site: Cultural resource management and archaeological research. *American Antiquity* 49:288–314.
1991a    The Goshen Paleoindian complex: New data for Paleoindian research. In *Clovis: Origins and Adaptations*, ed. R. Bonnichsen and K. L. Turnmire, 133–151. Corvallis, OR: Center for the Study of the First Americans.
1991b    *Prehistoric Hunters of the High Plains*. 2nd ed. San Diego: Academic Press.
1996a    Introduction. In *The Mill Iron Site*, ed. G. C. Frison, 1–13. Albuquerque: University of New Mexico Press.
1996b    *The Mill Iron Site*. Albuquerque: University of New Mexico Press.

Frison, G. C., and B. A. Bradley
1980    *Folsom Tools and Technology at the Hanson Site, Wyoming*. Albuquerque: University of New Mexico Press.
1999    *The Fenn Cache: Clovis Weapons and Tools*. Albuquerque: One Horse Land and Cattle Company.

Frison, G. C., C. V. Haynes Jr., and M. L. Larson
1996    Discussion and conclusions. In *The Mill Iron Site*, ed. G. C. Frison, 205–216. Albuquerque: University of New Mexico Press.

Frison, G. C., and D. J. Stanford
1982a    *The Agate Basin Site: A Record of the Paleoindian Occupation of the Northwestern High Plains*. New York: Academic Press.
1982b    Summary and conclusions. In Frison and Stanford, *The Agate Basin Site*, 361–370. New York: Academic Press.

Gallivan, M. D.
2002    Measuring sedentariness and settlement population: Accumulations research in the Middle Atlantic region. *American Antiquity* 67:535–557.

Giraldeau, L.-A., and T. Caraco
2000    *Social Foraging Theory*. Princeton, NJ: Princeton University Press.

Gosse, J. C., E. B. Evenson, J. Klein, B. Lawn, and R. Middleton
1995   Precise cosmogenic 10Be measurements in western North America: Support for a global Younger Dryas cooling event. *Geology* 23:877–880.

Gould, R. A.
1969   *Yiwara: Foragers of the Australian Desert*. New York: Charles Scribner and Sons.
1977   *Puntutjarpa Rockshelter and the Australian Desert Culture*. Anthropological Papers of the American Museum of Natural History, vol. 54, pt. 1. New York: American Museum of Natural History.

Gramly, R. M.
1993   *The Richey Clovis Cache*. Kenmore, NY: Persimmon Press.

Green, F. E.
1963   The Clovis blades: An important addition to the Llano complex. *American Antiquity* 29:145–165.

Greiser, S. T.
1985   *Predictive Models of Hunter-Gatherer Subsistence and Settlement Strategies on the Central High Plains*. Plains Anthropologist Memoir 20. [Iowa City, IA]: Plains Anthropologist.

Grinnell, G. B.
1972 [1923]   *The Cheyenne Indians: Their History and Ways of Life*. Vols. 1 and 2. Lincoln: University of Nebraska Press.

Gryba, E. M.
1985   Evidence of the fluted point tradition in Alberta. *Alberta Archaeological Review* 11:3–11.

Gunnerson, J. H.
1956   A fluted point site in Utah. *American Antiquity* 21:412–414.

Hall, C. T., and M. L. Larson, eds.
2004   *Aggregate Analysis in Chipped Stone*. Salt Lake City: University of Utah Press.

Hames, R.
1992   Time allocation. In *Evolutionary Ecology and Human Behavior*, ed. E. A. Smith and B. Winterhalder, 203–235. New York: Aldine de Gruyter.

Hartwell, W. T.
1995   The Ryan's site cache: Comparisons to Plainview. *Plains Anthropologist* 40:165–184.

Haury, E. W., E. B. Sayles, and W. W. Wasley
1959   The Lehner mammoth site, southeastern Arizona. *American Antiquity* 25 (1): 2–42.

Hawkes, K., K. Hill, and J. F. O'Connell
1982   Why hunters gather: Optimal foraging and the Ache of eastern Paraguay. *American Ethnologist* 9:379–398.

Hayden, B., N. Franco, and J. Spafford
1996   Evaluating lithic strategies and design criteria. In *Stone Tools: Theoretical Insights into Human Behavior*, ed. G. H. Odell, 9–49. New York: Plenum Press.

Haynes, C. V., Jr.
1991a  Clovis-Folsom-Midland-Plainview geochronology, climate change, and extinction. Paper presented at the 56th annual meeting of the Society for American Archaeology, New Orleans, Louisiana.
1991b  Geoarchaeological and paleohydrological evidence for a Clovis-Age drought in North America and its bearing on extinction. *Quaternary Research* 35: 438–450.
1992  Contributions of radiocarbon dating to the geochronology of the peopling of the New World. In *Radiocarbon after Four Decades*, ed. R. E. Taylor, A. Long, and S. Kra, 355–374. New York: Springer-Verlag.
1993  Clovis-Folsom geochronology and climatic change. In *From Kostenki to Clovis: Upper Paleolithic–Paleo-Indian Adaptations*, ed. O. Soffer and N. D. Praslov, 219–236. New York: Plenum Press.
Haynes, C. V., Jr., R. P. Beukens, A.J.T. Jull, and O. K. Davis
1992  New radiocarbon dates for some old Folsom sites: Accelerator technology. In *Ice Age Hunters of the Rockies*, ed. D. J. Stanford and J. S. Day, 83–100. Niwot: Denver Museum of Natural History and University Press of Colorado.
Hegmon, M.
2003  Setting theoretical egos aside: Issues and theory in North American archaeology. *American Antiquity* 68:213–243.
Hesse, P. P., J. W. Magee, and S. van der Kaars
2004  Late Quaternary climates of the Australian arid zone: A review. *Quaternary International* 118–119:87–102.
Hester, J. J.
1962  A Folsom lithic complex from the Elida site, Roosevelt County, N.M. *El Palacio* 69 (2): 92–113.
1972  Blackwater Locality No. 1: A Stratified, Early Man Site in Eastern New Mexico. Fort Burgwin Research Center, Southern Methodist University, Rancho de Taos, New Mexico.
Hester, J. J., and J. Grady
1977  Paleoindian social patterns on the Llano Estacado. In *Paleo-Indian Lifeways*, ed. E. Johnson, 78–96. Lubbock: West Texas Museum Association.
Hill, M. G.
1994  Subsistence strategies by Folsom hunters at Agate Basin, Wyoming: A taphonomic analysis of the bison and pronghorn assemblages. MA thesis, University of Wyoming.
2001  Paleoindian diet and subsistence behavior on the Northwestern Great Plains of North America. PhD diss., University of Wisconsin.
Hill, M. G., and F. Sellet
2000  Refinements of Folsom subsistence and technological organization at Agate Basin, Wyoming. Paper presented at the joint Midwest Archaeological and Plains Anthropological Conference, St. Paul, Minnesota.

Hiscock, P., and P. Veth
1991    Change in the Australian desert culture: A reanalysis of tulas from Puntutjarpa rockshelter. *World Archaeology* 22:332–345.
Hofman, J. L.
1990    Salt Creek, Recent evidence from the eastern Folsom margin in central Oklahoma. *Plains Anthropologist* 35:367–374.
1991    Folsom land use: Projectile point variability as a key to mobility. In *Raw Material Economies among Prehistoric Hunter-Gatherers*, 2 vols., ed. A. Montet-White and S. Holen, 285–303. Lawrence: University of Kansas.
1992    Recognition and interpretation of Folsom technological variability on the Southern Plains. In *Ice Age Hunters of the Rockies*, ed. D. J. Stanford and J. S. Day, 193–224. Niwot: Denver Museum of Natural History and University Press of Colorado.
1994    Paleoindian aggregations on the Great Plains. *Journal of Anthropological Archaeology* 13:341–370.
1995    Dating Folsom occupations on the Southern Plains: The Lipscomb and Waugh sites. *Journal of Field Archaeology* 22:421–437.
1999a   Unbounded hunters: Folsom bison hunting on the Southern Plains circa 10500 BP, the lithic evidence. In *Le Bison: Gibier et Moyen de Subsistance des Hommes du Paléolithique Aux Paléoindiens des Grandes Plaines*, ed. J.-P. Brugal, F. David, J. G. Enloe, and J. Jaubert, 383–415. Toulouse: Actes du Colloque International.
1999b   Folsom fragments, site types, and assemblage formation. In *Folsom Lithic Technology: Explorations in Structure and Variation*, ed. D. S. Amick, 122–143. Ann Arbor, MI: International Monographs in Prehistory.
Hofman, J. L., D. S. Amick, and R. O. Rose
1990    Shifting sands: A Folsom-Midland assemblage from a campsite in western Texas. *Plains Anthropologist* 35:221–253.
Hofman, J. L., and B. J. Carter
1991    The Waugh site: A Folsom-bison association in northwestern Oklahoma. In *A Prehistory of the Plains Border Region*, ed. B. J. Carter and P. A. Ward, 24–37. Stillwater: Department of Agronomy, Oklahoma State University.
Hofman, J. L., and E. Ingbar
1988    A Folsom hunting overlook in eastern Wyoming. *Plains Anthropologist* 33:337–350.
Holdaway, S., S. McPherron, and B. Roth
1996    Notched tool reuse and raw material availability in French Middle Paleolithic sites. *American Antiquity* 61:377–387.
Holliday, V. T.
1997    *Paleoindian Geoarchaeology of the Southern High Plains*. Austin: University of Texas Press.
2000    The evolution of Paleoindian geochronology and typology on the Great Plains. *Geoarchaeology* 15:227–290.

Holliday, V. T., E. Johnson, H. Haas, and R. Stuckenrath
1983   Radiocarbon ages from the Lubbock Lake site, 1950–1980: Framework for cultural and ecological change on the southern high plains. *Plains Anthropologist* 28:165–182.
1985   Radiocarbon ages from the Lubbock Lake site, 1981–1984. *Plains Anthropologist* 30:277–291.

Holliday, V. T., E. Johnson, and T. W. Stafford Jr.
1999   AMS radiocarbon dating of the type Plainview and Firstview (Paleoindian) assemblages: The agony and ecstasy. *American Antiquity* 64:444–454.

Howard, E. B.
1939   Folsom and Yuma points from Saskatchewan. *American Antiquity* 3:277–279.

Huckell, B. B., and C. V. Haynes Jr.
2003   The Ventana complex: New dates and new ideas on its place in early Holocene western prehistory. *American Antiquity* 68:353–371.

Hunt, A. P., and D. Tanner
1960   Early man sites near Moab, Utah. *American Antiquity* 26:110–117.

Ingbar, E. E.
1992   The Hanson Site and Folsom on the Northwestern Plains. In *Ice Age Hunters of the Rockies*, ed. D. J. Stanford and J. S. Day, 169–192. Niwot: Denver Museum of Natural History and University Press of Colorado.
1994   Lithic material selection and technological organization. In *The Organization of North American Chipped Stone Technologies*, ed. P. J. Carr, 45–56. Ann Arbor, MI: International Monographs in Prehistory.

Ingbar, E. E., and J. L. Hofman
1999   Folsom fluting fallacies. In *Folsom Lithic Technology: Explorations in Structure and Variation*, ed. D. S. Amick, 98–110. Ann Arbor, MI: International Monographs in Prehistory.

Irwin, H.
1971   Developments in early man studies in western North America. *Arctic Anthropology* 8 (2): 42–67.

Irwin, H. T., and H. M. Wormington
1970   Paleoindian tool types in the Great Plains. *American Antiquity* 35:24–34.

Irwin-Williams, C., H. Irwin, G. Agogino, and C. V. Haynes
1973   Hell Gap: Paleo-indian occupation on the High Plains. *Plains Anthropologist* 18:40–53.

Izett, G. A.
1968   *Geology of the Hot Sulphur Springs Quadrangle, Grand County, Colorado.* Geological Survey Professional Paper 586. Washington DC: U.S. Government Printing Office.

Jodry, M. A.
1998   The possible design of Folsom ultrathin bifaces as fillet knives for jerky production. *Current Research in the Pleistocene* 15:75–77.

Jodry, M. A., M. D. Turner, V. Spero, J. C. Turner, and D. Stanford
1996   Folsom in the Colorado high country: The Black Mountain site. *Current Research in the Pleistocene* 13:25-27.

Jodry, M.A.B.
1999   Folsom technological and socioeconomic strategies: Views from Stewart's Cattle Guard and the Upper Rio Grande Basin, Colorado. PhD diss., American University.

Johnson, E.
1987   *Lubbock Lake: Late Quaternary Studies on the Southern High Plains.* College Station: Texas A&M University Press.
1997   Late Quaternary bison utilization at Lubbock Lake, Southern High Plains of Texas. In *Southern Plains Bison Procurement and Utilization from Paleoindian to Historic*, ed. L. C. Bement and K. J. Buehler, 45-61. Plains Anthropologist Memoir 29. Lincoln, NE: Plains Anthropological Society.

Johnson, J. K.
1989   The utility of production trajectory modeling for regional analysis. In *Alternative Approaches to Lithic Analysis*, ed. D. O. Henry and G. H. Odell, 119-138. Washington DC: American Anthropological Association.

Judge, W. J.
1973   *Paleoindian Occupation of the Central Rio Grande Valley in New Mexico.* Albuquerque: University of New Mexico Press.

Kehoe, T.
1966   The distribution and implications of fluted points in Saskatchewan. *American Antiquity* 31:530-539.

Kelly, R. L.
1983   Hunter-gatherer mobility strategies. *Journal of Anthropological Research* 39:277-306.
1988   The three sides of a biface. *American Antiquity* 53:717-734.
1992   Mobility/sedentism: Concepts, archaeological measures, and effects. *Annual Review of Anthropology* 21:43-66.
1995   *The Foraging Spectrum.* Washington DC: Smithsonian Institution Press.

Kelly, R. L., and L. C. Todd
1988   Coming into the country: Early Paleoindian hunting and mobility. *American Antiquity* 53:231-244.

Kintigh, K. W.
1984   Measuring archaeological diversity by comparison with simulated assemblages. *American Antiquity* 49:44-54.

Kitcher, P.
1985   *Vaulting Ambition: Sociobiology and the Quest for Human Nature.* Cambridge, MA: MIT Press.

Kohler, T. A.
1978   Ceramic breakage rate simulation: Population size and the southeastern chiefdom. *Newsletter of Computer Archaeology* 14:1-13.

Kohler, T. A., and E. Blinman
1987 Solving mixture problems in archaeology: Analysis of ceramic materials for dating and demographic reconstruction. *Journal of Anthropological Archaeology* 6:1–28.

Kooyman, B., M. E. Newman, C. Cluney, M. Lobb, S. Tolman, P. McNeil, and L. V. Hills
2001 Identification of horse exploitation by Clovis hunters based on protein analysis. *American Antiquity* 66:686–691.

Kornfeld, M.
1988 The Rocky Folsom site: A small Folsom lithic assemblage from the northwestern plains. *North American Archeologist* 9:197–222.
1998 Folsom technology and subsistence. In *Early Prehistory of Middle Park: The 1997 Project and Summary of Paleoindian Archaeology*, ed. M. Kornfeld, 56–62. Technical Report, No. 15a. Laramie: Department of Anthropology, University of Wyoming.

Kornfeld, M., and G. C. Frison
2000 Paleoindian occupation of the high country: The case of Middle Park, Colorado. *Plains Anthropologist* 45:129–153.

Kornfeld, M., G. C. Frison, M. L. Larson, J. C. Miller, and J. Saysette
1999 Paleoindian bison procurement and paleoenvironments in Middle Park of Colorado. *Geoarchaeology* 14:655–674.

Kornfeld, M., G. C. Frison, and P. White
2001 Paleoindian occupation of Barger Gulch and the use of Troublesome formation chert. *Current Research in the Pleistocene* 18:32–34.

Krebs, J. R., and N. B. Davies, eds.
1984 *Behavioural Ecology: An Evolutionary Approach*. Oxford: Blackwell.

Kuhn, S. L.
1989 Hunter-gatherer foraging organization and strategies of artifact replacement and discard. In *Experiments in Lithic Technology*, ed. D. S. Amick and R. P. Mauldin, 33–47. Oxford: British Archaeological Reports.
1991 "Unpacking" reduction: Lithic raw material economy in the Mousterian of West-Central Italy. *Journal of Anthropological Archaeology* 10:76–106.
1994 A formal approach to the design and assembly of mobile toolkits. *American Antiquity* 59:426–442.
1995 *Mousterian Lithic Technology*. Princeton, NJ: Princeton University Press.

Larson, M. L.
1994 Toward a holistic analysis of chipped stone assemblages. In *The Organization of North American Chipped Stone Technologies*, ed. P. J. Carr, 57–69. Ann Arbor, MI: International Monographs in Prehistory.

Lee, R. B.
1979 *The !Kung San: Men, Women, and Work in a Foraging Society*. Cambridge: Cambridge University Press.

LeTourneau, P. D.
1998   The "Folsom Problem." In *Unit Issues in Archaeology*, ed. A. F. Ramenofsky, 52–73. Salt Lake City: University of Utah Press.
2001   Evidence of the role of bifacial cores in Folsom lithic technology. *Current Research in the Pleistocene* 18:36–39.

Lightfoot, K., and R. Jewett
1986   The shift to sedentary life: A consideration of the occupation duration of early Mogollon pithouse villages. In *Mogollon Variability*, ed. C. Benson and S. Upham. University Museum, Occasional Papers No. 15. New Mexico State University, Las Cruces.

Lightfoot, K. G., and R. A. Jewett
1984   The occupation duration of Duncan. In *The Duncan Project: A Study of the Occupation Duration and Settlement Pattern of an Early Mogollon Pithouse Village*, ed. K. G. Lightfoot. Anthropological Field Studies Number 6. Office of Cultural Resource Management, Department of Anthropology, Arizona State University, Tempe.

Long, A., and B. Rippeteau
1974   Testing contemporaneity and averaging radiocarbon dates. *American Antiquity* 39:205–214.

Love, C. M.
1997   Final report on the chert sources of the Western Geophysical Table Rock–Higgins 3-D Seismic Program, Southwest Wyoming. Unpublished manuscript on file at the University of Wyoming.

MacDonald, D. H.
1998a  Subsistence, sex, and cultural transmission in Folsom culture. *Journal of Anthropological Research* 17:217–239.
1998b  Subsistence, tool-use, and reproductive strategies of Northern Plains Folsom hunter-gatherers: A view from the Bobtail Wolf site, North Dakota. PhD diss., Washington State University.
1999   Modeling Folsom mobility, mating strategies, and technological organization in the Northern Plains. *Plains Anthropologist* 44:141–161.

MacDonald, D. H., and B. S. Hewlett
1999   Reproductive interests and forager mobility. *Current Anthropology* 40: 501–523.

Marks, A. E., J. Shokler, and J. Zilhão
1991   Raw material usage in the Paleolithic: The effects of local availability on selection and economy. In *Raw Material Economies among Prehistoric Hunter-Gatherers*, ed. A. Montet-White and S. Holen, 127–139. University of Kansas Publications in Anthropology 19. Lawrence: University Press of Kansas.

Mayer, J. H.
2002   Evaluating natural site formation processes in eolian dune sands: A case study from the Krmpotich Folsom site, Killpecker Dunes, Wyoming. *Journal of Archaeological Science* 29:1199–1211.

2003    Paleoindian geoarchaeology and paleoenvironments of the western Killpecker Dunes, Wyoming, U.S.A. *Geoarchaeology* 18:35–69.

Mayer, J. H., T. A. Surovell, N. M. Waguespack, M. Kornfeld, R. G. Reider, and G. C. Frison

2005    Paleoindian environmental change and landscape response in Barger Gulch, Middle Park, Colorado. *Geoarchaeology* 20:599–625.

McDonald, J. N.

1981    *North American Bison: Their Classification and Evolution*. Berkeley and Los Angeles: University of California Press.

Meltzer, D. J.

1988    Late Pleistocene human adaptations in eastern North America. *Journal of World Prehistory* 2:1–52.

1993    Is there a Clovis adaptation? In *From Kostenki to Clovis: Upper Paleolithic: Paleo-Indian Adaptations*, ed. O. Soffer and N. D. Praslov, 293–310. New York: Plenum Press.

2006a   Defining Folsom: Themes and variations. In *Folsom: New Archaeological Investigations of a Classic Paleoindian Bison Kill*, ed. D. J. Meltzer, 338–344. Berkeley and Los Angeles: University of California Press.

2006b   Folsom: From prehistory to history. In *Folsom: New Archaeological Investigations of a Classic Paleoindian Bison Kill*, ed. D. J. Meltzer, 295–307. Berkeley and Los Angeles: University of California Press.

Meltzer, D. J., and V. T. Holliday

2006    Geology, paleotopography, stratigraphy, and geochronology. In *Folsom: New Archaeological Investigations of a Classic Paleoindian Bison Kill*, ed. D. J. Meltzer, 112–153. Berkeley and Los Angeles: University of California Press.

Meltzer, D. J., and B. D. Smith

1986    Paleoindian and Early Archaic subsistence strategies in eastern North America. In *Foraging, Collecting, and Harvesting: Archaic Period Subsistence and Settlement in the Eastern Woodlands*, ed. S. Neusius, 3–31. Carbondale: Center for Archaeological Investigations, Southern Illinois University.

Meltzer, D. J., and L. C. Todd

2006    The faunal assemblage and bison bonebed taphonomy. In *Folsom: New Archaeological Investigations of a Classic Paleoinidian Bison Kill*, ed. D. J. Meltzer, 205–246. Berkeley and Los Angeles: University of California Press.

Meltzer, D. J., L. C. Todd, and V. T. Holliday

2002    The Folsom (Paleoindian) type site: Past investigations, current studies. *American Antiquity* 67:5–36.

Metcalfe, D., and K. R. Barlow

1992    A model for exploring the optimal trade-off between field processing and transport. *American Anthropologist* 94:340–356.

Miller, J. C.

1991    Lithic resources. In *Prehistoric Hunters of the High Plains*, ed. G. C. Frison, 449–476. San Diego: Academic Press.

Miller, J. C.
1996    Lithic sources in the Northwestern Plains. In *Archeological and Bioarcheological Resources of the Northern Plains*, ed. G. C. Frison and R. C. Mainfort, 41–49. Fayetteville: Arkansas Archeological Survey.

Miller, S. J.
1982    The archaeology and geology of an extinct megafauna/fluted point association at Owl Cave, the Wasden site, Idaho: A preliminary report. In *Peopling of the New World*, ed. J. E. Ericson, R. E. Taylor, and R. Berger, 81–95. Los Altos, CA: Ballena Press.

Mills, B. J.
1989    Integrating functional analyses of vessels and sherds through models of ceramic assemblage formation. *World Archaeology* 21:133–147.

Moore, J. H.
1987    *The Cheyenne Nation*. Lincoln: University of Nebraska Press.

Morrow, J. E., and T. A. Morrow
2002    Rummells-Maske revisited: A fluted point cache from east central Iowa. *Plains Anthropologist* 47:307–321.

Morrow, T. A.
1996    Bigger is better: Comments on Kuhn's formal approach to mobile tool kits. *American Antiquity* 61:581–590.

Morrow, T. A., and J. E. Morrow
1999    On the fringe: Folsom points and preforms in Iowa. In *Folsom Lithic Technology: Explorations in Structure and Variation*, ed. D. S. Amick, 65–81. Ann Arbor, MI: International Monographs in Prehistory.

Moss, E. H.
1986    Aspects of site comparison: Debitage samples, technology, and function. *World Archaeology* 18:116–133.

Munson, P. J.
1990    Folsom fluted projectile points east of the Great Plains and their biogeographical correlates. *North American Archaeologist* 11:255–272.

Nami, H. G.
1999    The Folsom biface reduction sequence: Evidence from the Lindenmeier collection. In *Folsom Lithic Technology: Explorations in Structure and Variation*, ed. D. S. Amick, 82–97. Ann Arbor, MI: International Monographs in Prehistory.

Naroll, R.
1962    Floor area and settlement population. *American Antiquity* 27:587–589.

Nash, S. E.
1996    Is curation a useful heuristic? In *Stone Tools: Theoretical Insights into Human Behavior*, ed. G. H. Odell, 81–99. New York: Plenum Press.

Naze, B.
1994    The Crying Woman site: A record of prehistoric human habitation in the Colorado Rockies. MA thesis, Colorado State University.

Naze, B. S.
1986  The Folsom occupation of Middle Park, Colorado. *Southwestern Lore* 52 (4): 1–32.

Nelson, M. C.
1991  The study of technological organization. In *Archaeological Method and Theory*, ed. M. B. Schiffer, 57–100. Tucson: University of Arizona Press.

O'Connell, J. F., and K. Hawkes
1981  Alyawara plant use and optimal foraging theory. In *Hunter-Gatherer Foraging Strategies*, ed. B. Winterhalder and E. A. Smith, 99–125. Chicago: University of Chicago Press.

Olszewski, D. I.
1989  Tool blank selection, debitage, and cores from Abu Hureyra I, northern Syria. *Paléorient* 15 (2): 29–37.

Parry, W. J.
1994  Prismatic blade technologies in North America. In *The Organization of North American Chipped Stone Technologies*, ed. P. J. Carr, 87–98. Ann Arbor, MI: International Monographs in Prehistory.

Parry, W. J., and R. L. Kelly
1987  Expedient core technology and sedentism. In *The Organization of Core Technology*, ed. J. K. Johnson and C. A. Morrow, 285–304. Boulder, CO: Westview Press.

Pascal, B.
1670  *Pensées de M. Pascal sur la réligion, et sur quelques autres sujets.* Paris: Guillaume Desprez.

Peterson, M. R.
2001  Folsom mobility and technological organization at the Krmpotich site: An analysis of the lithic artifact assemblage. MA thesis, University of Wyoming.

Polyak, V. J., J.B.T. Rasmussen, and Y. Asmerom
2004  Prolonged wet period in the southwestern United States through the Younger Dryas. *Geology* 32:5–8.

Prasciunas, M. M.
2004  Bifacial versus amorphous core technology: Experimental testing of differential flake tool production. MA thesis, University of Wyoming.

Rasic, J., and W. Andrefsky Jr.
2001  Alaskan blade cores as specialized components of mobile toolkits: Assessing design parameters and toolkit organization through debitage analysis. In *Lithic Debitage: Context, Form, Meaning*, ed. W. Andrefsky Jr., 61–79. Salt Lake City: University of Utah Press.

Reasoner, M. A., and M. A. Jodry
2000  Rapid response of alpine timberline vegetation to the Younger Dryas climate oscillation in the Colorado Rocky Mountains, U.S.A. *Geology* 28:51–54.

Reher, C. A.
1991  Large-scale lithic quarries and regional transport systems on the High Plains of eastern Wyoming. In *Raw Material Economies among Prehistoric*

*Hunter-Gatherers*, ed. A. Montet-White and S. Holen, 251–285. University of Kansas Publications in Anthropology 19. Lawrence: University Press of Kansas.

Reiss, D., L. Shaw, D. Eckles, and J. Hauff

1980     *Final Report of Archeological Investigations of Kerr-McGee Coal Corporations East Gillette Mine Permit Area, Campbell County, Wyoming*. Report on file at the Office of the Wyoming State Archeologist, Laramie, Wyoming.

Roberts, F.H.H., Jr.

1936     Further investigations at a Folsom campsite in northern Colorado. In *Explorations and Field-work of the Smithsonian Institution in 1935*, 69–74. Washington DC: Smithsonian Institution.

1943     A new site. *American Antiquity* 8 (3): 300.

1961     The Agate Basin complex. In *Homenaje a Pablo Martínez del Río en el XXV Aniversario de la Edición de Los Orígenes Americanos*, 125–132. Mexico City: Instituto Nacional de Antropología e Historia.

Rolland, N., and H. Dibble

1990     A new synthesis of Middle Paleolithic variability. *American Antiquity* 55: 480–499.

Root, M. J., ed.

2000     *The Archaeology of the Bobtail Wolf Site: Folsom Occupation of the Knife River Flint Quarry Area*. Pullman: Washington State University Press.

Root, M. J., J. D. William, and A. M. Emerson

2000     Stone tools and flake debris. In *The Archaeology of the Bobtail Wolf Site*, ed. M. J. Root, 223–308. Pullman: Washington State University Press.

Root, M. J., J. D. William, M. Kay, and L. K. Shifrin

1999     Folsom ultrathin biface and radial break tools in the Knife River Flint quarry area. In *Folsom Lithic Technology: Explorations in Structure and Variation*, ed. D. S. Amick, 144–168. Ann Arbor, MI: International Monographs in Prehistory.

Ross, A., T. Donnelly, and R. Wasson

1992     The peopling of the arid zone: Human-environment interactions. In *The Naïve Lands: Prehistory and Environmental Change in Australia and the Southwest Pacific*, ed. J. Dodson, 76–114. Melbourne: Longman Cheshire.

Sahlins, M.

1972     *Stone Age Economics*. Chicago: Aldine.

1976     *The Use and Abuse of Biology: An Anthropological Critique of Sociobiology*. Ann Arbor: University of Michigan Press.

Sanchez, M. G.

2001     A synopsis of Paleo-Indian archaeology in Mexico. *Kiva* 67:119–139.

Saunders, J. J.

1977     Lehner Ranch revisited. In *Paleo-Indian Lifeways*, ed. E. Johnson, 48–64. Lubbock: West Texas Museum Association.

Schiffer, M. B.

1975a     Archaeology as behavioral science. *American Anthropologist* 77:836–848.

Schiffer, M. B.
1975b   The effects of occupation span on site content. In *The Cache River Archaeological Project: An Experiment in Contract Archaeology*, ed. M. B. Schiffer and J. H. House, 265–269. Arkansas Archeological Survey, Publications in Archeology, Research Series, No. 8. Fayetteville: Arkansas Archaeological Survey.
1987    *Formation Processes of the Archaeological Record*. Albuquerque: University of New Mexico Press.
1988    The structure of archaeological theory. *American Antiquity* 53 (3): 461–485.
1996    Some relationships between behavioral and evolutionary archaeology. *American Antiquity* 61:643–662.
1999    Behavioral archaeology: Some clarifications. *American Antiquity* 64:166–168.

Schlanger, S. H.
1990    Artifact assemblage composition and site occupation duration. In *Perspectives on Southwestern Prehistory*, ed. P. E. Minnis and C. L. Redman, 103–121. Boulder, CO: Westview Press.

Schroedl, A. R.
2000    Reassessing the Silverhorn Folsom site in central Utah. *Current Research in the Pleistocene* 17:72–73.

Sellards, E. H., G. L. Evans, G. Meade, and A. D. Krieger
1947    Fossil bison and associated artifacts from Plainview, Texas. *Bulletin of the Geological Society of America* 58:927–954.

Sellet, F.
1999    A dynamic view of Paleoindian assemblages at the Hell Gap site, Wyoming: Reconstructing lithic technological systems. PhD diss., Southern Methodist University.
2001    A changing perspective on Paleoindian chronology and typology: A view from the Northwestern Plains. *Arctic Anthropology* 38 (2): 48–63.
2004    Beyond the point: Projectile point manufacture and behavioral inference. *Journal of Archaeological Science* 31:1553–1566.

Sellet, F., and M. Fosha
2000    Distribution of Folsom and Goshen artifacts in South Dakota. *Current Research in the Pleistocene* 2000:74–75.

Semaw, S.
2000    The world's oldest stone artefacts from Gona, Ethiopia: Their implications for understanding stone technology and patterns of human evolution between 2.6–1.5 million years ago. *Journal of Archaeological Science* 27:1197–1214.

Sheets, P. D., and G. R. Muto
1972    Pressure blades and total cutting edge. *Science* 175:632–634.

Shifrin, L. K., ed.
2000    *Young-Man-Chief (32DU95D): A Folsom, Late Plains Archaic, and Late Prehistoric Site: Final Report of Investigations*. Pullman: Washington State University Press.

Shott, M.
1986  Settlement mobility and technological organization: An ethnographic examination. *Journal of Anthropological Research* 42:15–51.

Shott, M. J.
1989  On tool-class use lives and the formation of archaeological assemblages. *American Antiquity* 54:9–30.
1994  Size and form in the analysis of flake debris: Review and recent approaches. *Journal of Archaeological Method and Theory* 1:69–110.
1996  An exegesis of the curation concept. *Journal of Anthropological Research* 52:259–280.

Shott, M. J., and P. Sillitoe
2004  Modeling use-life distributions in archaeology using New Guinea Wola ethnographic data. *American Antiquity* 69:339–355.

Smith, A. M.
1974  *Ethnography of the Northern Utes.* Albuquerque: Museum of New Mexico Press.

Smith, E. A.
1991  *Inujjamiut Foraging Strategies: Evolutionary Ecology of an Arctic Hunting Economy.* New York: Aldine de Gruyter.
2000  Three styles in the evolutionary analysis of human behavior. In *Adaptation and Human Behavior: An Anthropological Perspective,* ed. L. Cronk, N. Chagnon, and W. Irons, 27–46. New York: Aldine de Gruyter.

Spiess, A. E.
1984  Arctic garbage and New England Paleo-Indians: The single occupation option. *Archaeology of Eastern North America* 12:280–285.

Stafford, T. W., Jr., P. E. Hare, L. Currie, A.J.T. Jull, and D. J. Donahue
1991  Accelerator radiocarbon dating at the molecular level. *Journal of Archaeological Science* 18:35–72.

Stanford, D.
1999  Analysis and interpretation of Hell Gap hunting strategies at the Jones-Miller site. In *Le Bison: Gibier et Moyen de Subistance des Hommes Du Paléolithique Aux Paléoindiens des Grandes Plaines,* ed. J.-P. Brugal, F. David, J. G. Enloe, and J. Jaubert, 437–454. Toulouse: Actes du Colloque International.

Stanford, D., and F. Broilo
1981  Frank's Folsom Campsite. *Artifact* 19 (2): 1–11.

Stanford, D. J., and M. A. Jodry
1988  The Drake Clovis cache. *Current Research in the Pleistocene* 5:21–22.

Stephens, D. W., and J. R. Krebs
1986  *Foraging Theory.* Princeton, NJ: Princeton University Press.

Steward, J. H.
1938  *Basin-Plateau Aboriginal Socio-Political Groups.* Smithsonian Institution Bureau of American Ethnology Bulletin 20. Washington DC: Government Printing Office.

Stiger, M.
2006  A Folsom structure in the Colorado Mountains. *American Antiquity* 71:321–351.
Surovell, T. A.
2000  Early Paleoindian women, children, mobility, and fertility. *American Antiquity* 65:493–509.
2003a  The behavioral ecology of Folsom lithic technology. PhD diss., University of Arizona.
2003b  Simulating coastal migration in New World colonization. *Current Anthropology* 484:580–591.
Surovell, T. A., and N. M. Waguespack
In press  Human prey choice in the late Pleistocene and its relation to megafaunal extinctions. In *American Megafaunal Extinctions at the End of the Pleistocene*, ed. G. Haynes. New York: Springer.
Surovell, T. A., N. M. Waguespack, and M. Kornfeld
2003  A note on the functions of Folsom ultrathins. *Current Research in the Pleistocene* 20:75–77.
Surovell, T. A., N. M. Waguespack, M. Kornfeld, and G. C. Frison
2001  Barger Gulch Locality B: A Folsom site in Middle Park, Colorado. *Current Research in the Pleistocene* 18:58–60.
2003  *The First Five Field Seasons at Barger Gulch, Locality B, Middle Park, Colorado.* Technical Report No. 26, George C. Frison Institute of Archaeology and Anthropology, University of Wyoming, Laramie.
Surovell, T. A., N. M. Waguespack, J. H. Mayer, M. Kornfeld, and G. C. Frison
2005  Shallow site archaeology: Artifact dispersal, stratigraphy, and radiocarbon dating at Barger Gulch, Locality B, Middle Park, Colorado. *Geoarchaeology* 20:627–649.
Surovell, T. A., N. M. Waguespack, S. Richings-Germain, M. Kornfeld, and G. C. Frison
2000  *1999 Investigations at the Barger Gulch and Jerry Craig Sites, Middle Park, Colorado.* Technical Report No. 18a, George C. Frison Institute of Archaeology and Anthropology, University of Wyoming, Laramie.
2001  *The 2000 Field Season at Barger Gulch, Locality B, Middle Park, Colorado.* Technical Report No. 19c, George C. Frison Institute of Archaeology and Anthropology, University of Wyoming, Laramie.
Tankersley, K. B.
1994  The effects of stone and technology on fluted point morphometry. *American Antiquity* 59:498–510.
Taylor, R. E., C. V. Haynes Jr., and M. Stuiver
1996  Clovis and Folsom age estimates: Stratigraphic context and radiocarbon calibration. *Antiquity* 70:515–525.
Thompson, S. P., and M. Gardner
1998  *Calculus Made Easy.* New York: St Martin's Press.

Todd, L. C.
1991  Seasonality studies and Paleoindian subsistence strategies. In *Human Predators and Prey Mortality*, ed. M. C. Stiner, 217–238. Boulder, CO: Westview Press.

Todd, L. C., J. L. Hofman, and C. B. Schultz
1990  Seasonality of the Scottsbluff and Lipscomb bison bonebeds: Implications for modeling Paleoindian subsistence. *American Antiquity* 55:813–827.
1992  Faunal analysis and Paleoindian studies: A reexamination of the Lipscomb bison bonebed. *Plains Anthropologist* 37:137–165.

Todd, L. C., D. J. Rapson, and J. L. Hofman
1996  Dentition studies of the Mill Iron and other early Paleoindian bison bonebed sites. In *The Mill Iron Site*, ed. G. C. Frison, 145–175. Albuquerque: University of New Mexico Press.

Torrence, R.
1983  Time budgeting and hunter-gatherer technology. In *Hunter-Gatherer Economy in Prehistory*, ed. G. Bailey, 11–22. Cambridge: Cambridge University Press.
1989a Tools as optimal solutions. In *Time, Energy, and Stone Tools*, ed. R. Torrence, 1–6. Cambridge: Cambridge University Press.
1989b *Time, Energy, and Stone Tools*. Cambridge: Cambridge University Press.
1989c Retooling: Towards a behavioral theory of stone tools. In *Time, Energy, and Stone Tools*, ed. R. Torrence, 57–66. Cambridge: Cambridge University Press.

Tunnell, C.
1977  Fluted projectile point production as revealed by lithic specimens from the Adair-Steadman site in northwest Texas. In *Paleo-Indian Lifeways*, ed. E. Johnson, 140–168. Lubbock: West Texas Museum Association.

Ugan, A., J. Bright, and A. Rogers
2003  When is technology worth the trouble? *Journal of Archaeological Science* 30:1315–1329.

Varien, M. D., and B. J. Mills
1997  Accumulations research: Problems and prospects for estimating site occupation span. *Journal of Archaeological Method and Theory* 4:141–191.

Varien, M. D., and J. M. Potter
1997  Unpacking the discard equation: Simulating the accumulation of artifacts in the archaeological record. *American Antiquity* 62:194–213.

Waguespack, N. M.
2003  Clovis hunting and the organization of subsistence labor. PhD diss., University of Arizona.

Waguespack, N. M., and T. A. Surovell
2003  Clovis hunting strategies, or how to make out on plentiful resources. *American Antiquity* 68:333–352.

Waguespack, N. M., T. A. Surovell, M. Kornfeld, and G. C. Frison
2002   *The 2001 Field Season at Barger Gulch, Locality B, Middle Park, Colorado.* Technical Report No. 20, George C. Frison Institute of Archaeology and Anthropology, University of Wyoming, Laramie.

Waguespack, N. M., T. A. Surovell, and J. P. Laughlin
2006   *The 2004 and 2005 Field Seasons at Barger Gulch, Locality B (5GA195).* Technical Report No. 41, George C. Frison Institute of Anthropology and Archaeology, Department of Anthropology, University of Wyoming, Laramie.

Weiss, K. M., R. F. Ferrell, and C. J. Hanis
1984   A New World syndrome of metabolic diseases with a genetic and evolutionary basis. *Yearbook of Physical Anthropology* 27 (S5): 153–178.

Wendorf, F., and T. H. Thomas
1951   Early man sites near Concho, Arizona. *American Antiquity* 17:107–114.

White, P. M.
1999   Getting the high altitude stone: Lithic technology at the Barger Gulch site (5GA195), Middle Park, Colorado. MA thesis, University of Wyoming.

Whitelaw, T.
1991   Some dimensions of variability in the social organization of community space among foragers. In *Ethnoarchaeological Approaches to Mobile Campsites*, ed. C. S. Gamble and W. A. Boismier, 131–188. Ann Arbor, MI: International Monographs in Prehistory.

Whitley, D. S., and R. I. Dorn
1993   New perspectives on the Clovis vs. pre-Clovis controversy. *American Antiquity* 58 (4): 626–647.

Whittaker, J. C.
1994   *Flintknapping: Making and Understanding Stone Tools.* Austin: University of Texas Press.

Wiessner, P.
1974   A functional estimator of population from floor area. *American Antiquity* 39:343–350.
1983   Style and social information in Kalahari San projectile points. *American Antiquity* 48:253–276.

Wilke, P. J., J. J. Flenniken, and T. L. Ozbun
1991   Clovis technology at the Anzick site, Montana. *Journal of California and Great Basin Anthropology* 13:242–272.

William, J. D., ed.
2000   *The Big Black Site (32DU955C): A Folsom Complex Workshop in the Knife River Flint Quarry Area, North Dakota.* Pullman: Washington State University Press.

Wilmsen, E. M., and F.H.H. Roberts Jr.
1984   *Lindenmeier, 1934–1974: Concluding Report on Investigations.* Smithsonian Contributions to Anthropology, No. 24. Washington DC: Smithsonian Institution Press.

Wilmsen, E. N.
1970  *Lithic Analysis and Cultural Inference: A Paleoindian Case.* Anthropological Papers of the University of Arizona, No. 16. Tucson: University of Arizona Press.

Wilson, E. O.
1994  *Naturalist.* Washington DC: Island Press.

Winterhalder, B.
1981  Foraging strategies in the boreal forest: An analysis of Cree hunting and gathering. In *Hunter-Gatherer Foraging Strategies*, ed. B. Winterhalder and E. A. Smith, 66–98. Chicago: University of Chicago Press.

Winterhalder, B., and C. Goland
1993  On population, foraging efficiency, and plant domestication. *Current Anthropology* 34:710–715.

Winterhalder, B., and E. A. Smith
1992  Evolutionary ecology and the social sciences. In *Evolutionary Ecology and Human Behavior*, ed. E. A. Smith and B. Winterhalder, 3–23. New York: Aldine de Gruyter.
2000  Analyzing adaptive strategies: Human behavioral ecology at twenty-five. *Evolutionary Anthropology* 9 (2): 51–72.

Woodburn, J.
1968  Stability and flexibility in Hadza residential groupings. In *Man the Hunter*, ed. R. B. Lee and I. DeVore, 103–110. Chicago: Aldine.

Wormington, H. M.
1957  *Ancient Man in North America.* Denver Museum of Natural History, Popular Series No. 4, Denver, Colorado.

Wyckoff, D. G.
1996  The Westfahl and Engle bifaces: Isolated finds of large bifaces on the Southern Plains. *Plains Anthropologist* 41 (157): 287–296.

Yellen, J. E.
1977  *Archaeological Approaches to the Present: Models for Reconstructing the Past.* New York: Academic Press.

# Index

accumulations research: review of, 61–67
Adair-Steadman site, 60, 224, 230
Agate Basin site: artifact density of, 103–7; artifact photographs from, 53; bison butchery at, 36–37; core density at, 190–91; cores from, 154; dating of, 30; debitage attributes from, 182–85; debitage:nonlocal tools at, 85, 86, 88, 89; distribution of lithic raw material sources at, 135; flake tool typology at, 43; lithic surplus at, 123–27; local:nonlocal raw materials, 80, 81, 85, 86, 88, 89; occupation history of, 96, 103–9, 223; Occupation Span Index of, 103–7; overview of, 50–53; photograph of, 52; retouch intensity of tools from, 195–200; seasonality of, 98; selectivity of blanks in tool production at, 205–10
Ahler, Stanley, 16
Alibates silicified dolomite, 39, 170
Allen site, 174
Alyawara, 100–101
Amick, Dan, 37, 38, 59–60, 227
Andrefsky, William, 153
Anzick site, 171
artifact density: vs. Core Reduction Index, 185; model of, 68–70, 105; Puntutjarpa shelter, of, 93, 101–3; reoccupation of sites, and, 100–9; of sites of study sample, 56, 60, 103–7; and use duration, 66

Bamforth, Douglas: on cores, 153, 174–75; on lithic curation, 12–13; on projectile points, 41; on raw material availability, 188
Barger Gulch Locality B: artifact density of, 103–7, 222–23; artifact photographs from, 47; core density at, 190, 191; cores from, 41, 154, 175; cost of lithic raw material procurement at, 135–36; dating of, 30; debitage attributes from, 182–85; debitage:nonlocal tools at, 85, 86, 88, 89; distribution of lithic raw material sources at, 133–35; lithic surplus at, 123–27, 133–37, 139–40; local:nonlocal raw materials at, 80, 81, 85–89; occupation history of, 96, 98, 103–9, 222–23; Occupation Span Index of, 103–7; overview of, 45–47; photograph of, 46; raw material use from, 60; retouch intensity on tools from, 195–200; selectivity of blanks in tool production at, 205–10; ultrathin biface from, 42
Barlow, K. Renee, 13–15, 176
Beaton, Jack, 39
Becker, Mark, 153, 174–75
behavioral ecology, 4–5, 7–9, 219–20

267

268  *Index*

bifaces: design of, 157–76, 221; distinguished from cores, 153; flake production from, 180–187; frequencies of in study sample, 81; and Paleoindian archaeologists, 142–43; photographs of, 47, 51, 53; traditional views of in Paleoindian archaeology, 42–43. *See also* bifacial cores; bifacial core tools; bifacial thinning flakes; bifacial tools; projectile points; ultrathin bifaces
bifacial cores: design of, 159–61, 167–76; transport cost of, 160; transport efficiency of, 160–61; utility of, 159
bifacial core tools: design of, 162–76; transport cost of, 162; transport efficiency of, 162–68; utility of, 162
bifacial thinning flakes: blanks for tools, as, 147–50, 171; density of, 222–23; discard rate of, 184–187, 191–210; modeled production of, 180–87; relative frequency of, 181–87
bifacial tools: design of, 158–60, 167–76; transport cost of, 159; transport efficiency of, 159–60; utility of, 159
Big Black site, 154
Binford, Lewis, 9–13, 128, 140, 200–201
*Bison antiquus*, 27, 33–38
Black Forest silicified wood, 46, 155
Blackwater Draw site, 41, 154, 170–71
Blinman, Eric, 64
Bobtail Wolf site: bone preservation at, 98; cores from, 41, 154, 175; lithic raw materials in, 39, 60; measures of mean per capita occupation span, of, 87–89; occupation history, 96; site function of, 224; ultrathin bifaces from, 42
Boldurian, Anthony, 177
Bonfire shelter, 35
Bordes, François, 43
Bradley, Bruce: on Goshen point morphology, 29; on Hanson site, 41, 43, 59–60, 89–90, 227

Brantingham, P. Jeffrey, 16–18
Bush, George W., 110–12
Byers, David, 37

caching, 138–39, 172
Carter/Kerr-McGee site: artifact density of, 103–7, 222–23; artifact photographs from, 55; core density of, 190, 191; cores from, 154; debitage attributes from, 182–85; debitage: nonlocal tools from, 85, 86, 88, 89; distribution of lithic raw material sources at, 135; Goshen/Clovis component of, 29, 31; lithic surplus at, 123–27; local:nonlocal raw materials, 80, 81, 85, 86, 88, 89; occupation history of, 96, 103–9, 222–23; Occupation Span Index of, 103–7; overview of, 53–56; photograph of, 54; retouch intensity on tools from, 195–200; selectivity of blanks in tool production at, 205–10
Casti, John, 213
Cattle Guard site, 154–55
central place foraging model: in lithic raw material procurement, 128–33; and mobility, 97, 216
channel flakes: frequencies of in study sample, 81; photographs of, 51, 55
Clarke, David, 213–14, 215, 219–20
Clarke Effect, 63
Clovis complex: at Agate Basin site, 50; bifaces from caches, 171; component at Carter/Kerr-McGee, 31, 54; general adaptation in, 229; Lehner site occupation history, 99; in mobility and reproduction, xiv; prey choice in, 34–35
Clovis site, 41, 154, 170–71
conservation of lithic raw material, 42, 44, 226–27
constraints in behavioral ecology, 7–9, 220–21

## Index

consumption rate, 113–21
Cooper site, 35, 36, 39, 87–89
core reduction flakes: as blanks for tools, 147–50, 174–76; discard rate of, 184–87, 191–210; on nonlocal raw materials, 156–57; production of, 180–87; relative frequency of, 181–87, 228
Core Reduction Index, 184–85
cores: as components of mobile toolkits, 150–57, 174–76; density of, 190–91, 199, 207; distinguished from bifaces, 153; flake production from, 180–87; frequencies of in study sample, 81; raw material of, 154–56; in tool production, 147–50, 174, 200–201; traditional views of in Paleoindian archaeology, 41–42; transport cost of, 150; transport efficiency of, 151–53; utility of, 150–51. *See also* bifacial cores; core reduction flakes
Cotter, John, 177
Crabtree, Donald, 172
curation, 10–13, 142, 231
currency in behavioral ecology, 7–9, 220–21

Dakota Formation quartzite, 48, 53
David, Nicholas, 61–63
d:e ratio. *See* lithic procurement, embedded
debitage: attributes of, 149, 156–57, 182, 184; definition of, 177; discard of, modeled, 177–212; frequencies of in study sample, 81; local production of, 82; reduction stage of, 157; in tool production, 147–50; use-life of, 178–79. *See also* bifacial thinning flakes; core reduction flakes
debitage:nonlocal tool ratio: definition of, 82–83; model of, 82–96; in Paleoindian sites, 83–90; at Puntutjarpa rockshelter, 93–95; and site reoccupation, 100–106
decision variable, in behavioral ecology, 7–9, 220–21
density. *See* artifact density
direct procurement. *See* lithic procurement, direct
discard rate: and artifact frequencies, 68; of debitage, 184–212; probabilistic nature of, 75; ratios of artifact types and, 71–74
Donohue, James, 32

Edwards Plateau chert, 38–39, 170
Elida site, 154–55
embedded procurement. *See* lithic procurement, embedded
endscrapers: as components of mobile toolkits, 146–47; morphology of, 43; photographs of, 47, 51, 55; size of, 146–47
evolutionary ecology. *See* behavioral ecology
expediency, 11, 137–38, 142–43, 231

Fagerström, Torbjörn, 218
field processing model, 13–15, 176
fitness, in behavioral ecology, 7–9
flake tools: vs. cores in mobile toolkits, 150–57, 174; design of, 15–18, 143–50; frequencies of in study sample, 81; functional efficiency of, 201–5, 220; manufacture of, 147–50; photographs of, 47, 49, 51, 53, 55; retouch intensity of, 194–200; selectivity of blanks in production of, 200–210; thickness of, 145–46, 147–49; traditional view of in Paleoindian archaeology, 42–43; transport cost of, 143–50; transport efficiency of, 144–45; use-life of, 191–200; utility of, 143–50
flexible technologies, 142, 231

fluting. *See* projectile points, fluting of
Folsom complex: age of, 29–30; geographic distribution of, 25–27; projectile point morphology in, 43
Folsom site, 30, 35, 36, 39
formal models: general considerations, 2, 19–21, 213–21
formal technologies, 137–38
founding curate set, 74–78
Frank's biface, 170–71
Frison, George: on Agate Basin site, 135; on Goshen complex, 27, 29, 31; on Hanson site, 41, 59–60, 89–90, 227

Gallivan, Martin, 64–67, 99
Geib, Phil, 16
Ghost site, 29
goals in behavioral ecology, 7–9
Gore, Albert, 110–12
Goshen complex: age of, 31–33; defined, 27; geographic distribution of, 28–29
Gould, Richard, 92–96, 101, 121

Hanson site: cores from, 41; debitage from, 174, 228; measures of mean per capita occupation span of, 87–89; occupation history of, 40, 58–60, 89–90, 96, 99, 108, 227–29; seasonality of, 98; site function of, 59–60, 224
Hartville chert, 53, 56
Haury, Emil, 99, 177
Haynes, C. Vance, Jr., 27, 30
Hay Springs site, 29
heavy-duty tools, 175–76
Hegmon, Michelle, 219
Hell Gap site: Agate Basin component from, 37; cores from, 154; Folsom component from, 60; Goshen component from, 27, 29, 30, 31
Hester, James, 155

high technology forager model, 43
Hill, Matthew G., 36, 52
Hiscock, Peter, 121
Hofman, Jack, 33, 39, 41, 170
Holliday, Vance, 30
hunting: in Folsom/Goshen contexts, 33–38

Indian Creek site, 34
Ingbar, Eric: on Folsom mobility as seen from Hanson site, 40, 58–60; on Hanson site occupation span, 89–90, 229; on Hanson site reoccupation, 99, 227
Irwin, Henry, 27, 43

Jewett, Roberta, 64–67, 99
Jim Pitts site, 29–31, 32
Jodry, Pegi, 33–34, 36, 42, 155
Johnson, Eileen, 34

Kelly, Robert: on bifaces, 42; on core reduction, 137–39; on hunter-gatherer mobility, 96–97; on lithic expediency, 188, 209; on models of mobility and technology, 2; on Paleoindian mobility, 37, 43, 221
Kohler, Timothy, 64
Krieger, Alex, 27, 29
Krmpotich site: artifact density of, 103–7, 223; artifact photographs from, 51; core density at, 190, 191; cores from, 154; cost of lithic raw material procurement at, 135–36; debitage attributes from, 182–85; debitage: nonlocal tools at, 85, 86, 88, 89; distribution of lithic raw material sources at, 134–35; lithic surplus at, 123–27; local:nonlocal raw materials at, 81, 85, 86, 88, 89; occupation history of, 96, 103–9, 223; Occupation Span Index of, 103–7; overview of, 48–51; photograph of, 50; raw

material use at, 50, 80; retouch intensity on tools from, 195–200; selectivity of blanks in tool production at, 205–10
Kuhn, Steven: on functional efficiency of tool blanks, 201; on lithic provisioning, 138–39; on model of mobile toolkits, 15–18, 142–59, 176; on Monty Hall problem, 214–15; on raw material availability, 188, 209

Lake Ilo sites. See Big Black site; Bobtail Wolf site
lake levels: Australian, 94–95
Lake Theo site: cores from, 154; measures of mean per capita occupation span for, 87–89; occupation history of, 96; and seasonal mobility, 39
Lehner site, 99
LeTourneau, Phillippe, 24, 41
Let's Make a Deal, 214–17
Levallois reduction, 16–18, 200
Lightfoot, Kent, 64–67, 99
Lindenmeier site: cores from, 154–55; debitage treatment in excavation at, 177; Folsom subsistence at, 34; raw material use at, 60, 224; tool blanks at, 41
Lipscomb site, 30, 35, 36, 39
lithic caching, 138–39, 172
lithic consumption rate, 113–21
lithic curation, 10–13, 142, 231
lithic expediency, 11, 137–38, 142–43, 231
lithic procurement: direct, 113–19, 128–33; embedded, 113–19, 128–33; rate of, 113–19; surplus creation in, 110–41
lithic shortfall, 113–19
lithic surplus: core transport and, 174; definitions of, 121–22, 123–24; model of, 110–41; optimal size of, 116–19, 220–21; Parry and Kelly model of, 137–39; at Puntutjarpa rockshelter, 119–24, 139; and raw material availability, 188–91, 210–11, 220; at sites of study sample, 123–27; as wasteful behavior, 226–27
local artifacts. See local:nonlocal raw material ratio
local:nonlocal raw material ratio: definition of, 78; model of, 74–78, 85; at Puntutjarpa rockshelter, 93–95; and site reoccupation, 100–106; at sites in study sample, 81, 85, 86, 88
logistical mobility: in raw material procurement, 128–33
Lohr, E., 177
Lower Twin Mountain site, 29
Lubbock Lake site, 30, 34, 35

MacDonald, Douglas, 39–40
maintainable technologies, 143, 231
marginal value theorem: applied to flake tools, 191–94, 221; applied to residential mobility, 91–92, 97
mean per capita occupation span: definition of, 67; in kill sites, 86–88; and lithic surplus creation, 113–27; models of, 70–98; at Puntutjarpa rockshelter, 93–96; in study sample, 78–81, 83–90. See also occupation span; Occupation Span Index
Meltzer, David, 36, 229
Metcalfe, Duncan, 13–15, 176
Middle Park, Colorado, 29, 45–48, 98
Mill Iron site: bison hunting at, 35; dating of, 32; and definition of Goshen, 27; measures of mean per capita occupation span of, 87–89; occupation history of, 96
mobile toolkits: definition of, 143; in Folsom/Goshen contexts, 170–76; Kuhn model of, 15–18, 142–59, 176; models of, 142–76
mobility. See logistical mobility; occupation span; residential mobility

Monty Hall problem, 214–17
Mountaineer site, 30
Munson, Patrick, 34

narrative modeling: in anthropology, 2–3, 218–20
Nelson, Margaret, 9, 187
nonlocal artifacts. *See* local:nonlocal raw material ratio

occupation intensity: definition of, 67
occupation span: and debitage discard, 178–87; definition of, 67; estimates of actual, 106–7, 222–23; and interpretations of Hanson site, 61–63; and lithic surplus creation, 113–39; models of, 68–78; raw material availability and, 188–91, 210–11; and reoccupation, 99–109. *See also* mean per capita occupation span; occupation intensity; Occupation Span Index; residential mobility
Occupation Span Index (OSI): vs. core density, 190–91; vs. Core Reduction Index, 185; definition of, 101–2; vs. lithic surplus size, 122–27; problem of variable discard rates and, 211–12; at Puntutjarpa rockshelter, 93, 101–3; vs. Retouch Intensity Index, 199; of sites of study sample, 103–6. *See also* mean per capita occupation span
O'Connell, James, 200–201
OSI. *See* Occupation Span Index
overlap: of sites, 108–9

Parry, William, 137–39, 188, 209
Phosphoria formation chert, 55, 59, 90, 228–29
Plainview complex, 27–28
platform attributes: of tools and debitage, 148–49, 156–57
Pleistocene extinctions, 34
Potter, James, 63–64

Prasciunas, Mary, 171, 194
preforms. *See* projectile point preforms
procurement. *See* lithic procurement
projectile point preforms, 47, 51, 55, 170, 172. *See also* projectile points, fluting of
projectile points: emphasis on in Paleoindian archaeology, 40–41; fluting of, 16, 25, 42, 44, 172; frequencies of in study sample, 81; life history of, 42, 170; in lithic surplus model, 115–21, 140; morphology of, 25, 172; photographs of, 47, 49, 51, 53, 55; spatial distribution of Folsom and Goshen, 27–29
provisioning people, 138–39
provisioning places, 138–39
Puntutjarpa rockshelter: artifact density of, 93, 101–3; lithic raw material use at, 92–95; lithic surplus at, 119–24, 139; measures of mean per capita occupation span at, 93–96; Occupation Span Index at, 93, 101–3; overview of, 92; water source at, 92

quarry sites, 58–59, 96, 223–26

radial break tools, 43
Rainy Day model. *See* lithic surplus
raw material: availability of, 12–13, 187–91, 203, 210–11, 220–21; conservation of, 42, 44, 226–27; transport of, 40
Reiss, David, 135
reliable technologies, 142, 231
reoccupation. *See* site reoccupation
residential mobility: cold season, 97–98, 222–23; Folsom and Goshen contexts, 96–97; general discussion of, 60–61; and raw material availability, 188–89, 210–11, 220; and site function, 224–26. *See also* occupation span; Occupation Span Index

residential stability, 64–67
retouch intensity, 194–200
Roberts, Frank H. H., 34, 155, 177
Root, Matthew, 42

Schiffer, Michael, 62–63, 71, 74, 82, 178–81
Schlanger, Sarah, 63
selectivity of blanks in tool production, 200–210
Sellet, Frederíc, 31, 32, 52
Shifting Sands site, 39
site area: model of, 233–36
site function, 223–26
site reoccupation, 58–74, 99–109
site types. *See* site function
Smith, Eric Alden, 21
sociobiology, 4–5. *See also* behavioral ecology
Spanish Diggings site, 224
spatial overlap of occupations, 108–9
Spearfish Formation chert, 53, 55
spherical cow problem, 20–21, 211–12
Spiess, Arthur, 43
Stafford, Thomas, 33
Stewart's Cattle Guard site, 154–55
surplus. *See* lithic surplus

technological organization, 9–10, 231
Todd, Larry, 36, 37, 43, 97–98, 221
toolkit size: optimal, 74–78
tools. *See* bifaces; bifacial tools; flake tools; heavy-duty tools; projectile points
Torrence, Robin, 1, 19
transport. *See* raw material, transport of
transported artifacts. *See* local:nonlocal raw material ratio
transport efficiency: of stone tools, 142–76
Troublesome Formation chert, 45, 47, 80–81, 133–34
typology: in divisions of surplus and consumed lithic raw materials, 121–22; of flake tools, 43; of projectile points, 24–25, 27–29

ultrathin bifaces: at Agate Basin site, 52; in Folsom technology, 42, 170; function of, 42; photographs of, 47, 52; and transport efficiency, 172–73
Upper Twin Mountain site: artifact density of, 103–7, 223; artifact photographs from, 49; bison hunting at, 35; core density at, 190, 191; cores from, 154; dating of 31, 32–33; debitage attributes from, 182–85; debitage:nonlocal tools, 85, 86–87, 88, 89; and definition of Goshen, 29; distribution of lithic raw material sources at, 135; lithic surplus at, 123–27; local:nonlocal raw materials at, 80–81, 85, 86–87, 88, 89; occupation history of, 96, 103–9, 223; Occupation Span Index of, 103–7; overview of, 46–49, 146–49; photograph of, 48; retouch intensity on tools from, 195–200; selectivity of blanks in tool production at, 205–10
use duration (of sites), 64–67
use-life (of artifacts): of bifaces, 42; of flake tools, 191–200; and lithic curation, 10–12; lithic raw material conservation and, 43–44; of projectile points, 16; as variable, 62–63, 71–72, 78, 82–83

Varien, Mark, 63–64
versatile technologies, 142, 231
Veth, Peter, 121

Waguespack, Nicole, 35
Waugh site, 30, 35
Wilmsen, Edwin, 34, 43, 155
Winterhalder, Bruce, 21
Wormington, Marie, 43

Yellen, John, 233–34
Younger Dryas, 30, 34

## About the Author

Todd A. Surovell is an assistant professor of anthropology at the University of Wyoming. After earning a bachelor's degree in anthropology and zoology in 1995 at the University of Wisconsin, Madison, he went on to earn his doctorate at the University of Arizona in 2003. He has participated in archaeological research in the United States, Denmark, and Israel. Focusing on issues surrounding the colonization of the Americas, lithic technology, and geoarchaeology, his work applies formal mathematical models and computer simulation to archaeological problems within a behavioral ecological framework. Integrating mathematical and archaeological approaches to the study of prehistoric hunter-gatherer decision making, Surovell has an extensive publication record addressing hunting strategies, demography, and archaeometry. Recent publications explore the predator-prey relationship between prehistoric hunters and their prey, and the spatial organization of activities at the Barger Gulch site. Surovell has spent the last ten years excavating Barger Gulch, a Folsom-aged campsite in Colorado, and has built upon his experiences with Folsom lithic technology to develop a novel and quantified approach to the archaeological interpretation of stone tools.

www.ingramcontent.com/pod-product-compliance
Lightning Source LLC
Chambersburg PA
CBHW031409290426
44110CB00011B/317